CAMBRIDGE LIBRARY COLLECTION

Books of enduring scholarly value

Botany and Horticulture

Until the nineteenth century, the investigation of natural phenomena, plants and animals was considered either the preserve of elite scholars or a pastime for the leisured upper classes. As increasing academic rigour and systematisation was brought to the study of 'natural history', its sub-disciplines were adopted into university curricula, and learned societies (such as the Royal Horticultural Society, founded in 1804) were established to support research in these areas. A related development was strong enthusiasm for exotic garden plants, which resulted in plant collecting expeditions to every corner of the globe, sometimes with tragic consequences. This series includes accounts of some of those expeditions, detailed reference works on the flora of different regions, and practical advice for amateur and professional gardeners.

An Account of the Foxglove, and Some of Its Medical Uses

In 1775, the physician and botanist William Withering (1741–99) was informed of a folk cure for dropsy that had as its active ingredient the plant foxglove (*Digitalis purpurea*). Ten years later, after thorough trials on more than 150 patients, Withering published this monograph on the medicinal applications of the plant, not least to keep less experienced doctors from administering it to patients without the proper caution, given the plant's toxicity. Withering was the first doctor to employ foxglove as a remedy for congestive heart failure, which is now the primary disease treated by foxglove-derived pharmaceuticals, and the results from his trials broadly reflect those produced by modern physicians. Withering's first major publication, *A Botanical Arrangement of All the Vegetables Naturally Growing in Great Britain* (1776), which includes observations on the medicinal applications of British plants, is also reissued in this series.

An Account of
the Foxglove,
and Some of Its Medical Uses

With Practical Remarks on Dropsy
and Other Diseases

WILLIAM WITHERING

CAMBRIDGE
UNIVERSITY PRESS

CAMBRIDGE
UNIVERSITY PRESS

University Printing House, Cambridge, CB2 8BS, United Kingdom

Cambridge University Press is part of the University of Cambridge.
It furthers the University's mission by disseminating knowledge in the pursuit of
education, learning and research at the highest international levels of excellence.

www.cambridge.org
Information on this title: www.cambridge.org/9781108075862

© in this compilation Cambridge University Press 2014

This edition first published 1785
This digitally printed version 2014

ISBN 978-1-108-07586-2 Paperback

Selected botanical reference works available in the
CAMBRIDGE LIBRARY COLLECTION

al-Shirazi, Noureddeen Mohammed Abdullah (compiler), translated by
Francis Gladwin: *Ulfáz Udwiyeh, or the Materia Medica* (1793)
[ISBN 9781108056090]

Arber, Agnes: *Herbals: Their Origin and Evolution* (1938)
[ISBN 9781108016711]

Arber, Agnes: *Monocotyledons* (1925) [ISBN 9781108013208]

Arber, Agnes: *The Gramineae* (1934) [ISBN 9781108017312]

Arber, Agnes: *Water Plants* (1920) [ISBN 9781108017329]

Bower, F.O.: *The Ferns (Filicales)* (3 vols., 1923–8) [ISBN 9781108013192]

Candolle, Augustin Pyramus de, and Sprengel, Kurt: *Elements of the Philosophy
of Plants* (1821) [ISBN 9781108037464]

Cheeseman, Thomas Frederick: *Manual of the New Zealand Flora*
(2 vols., 1906) [ISBN 9781108037525]

Cockayne, Leonard: *The Vegetation of New Zealand* (1928)
[ISBN 9781108032384]

Cunningham, Robert O.: *Notes on the Natural History of the Strait of Magellan
and West Coast of Patagonia* (1871) [ISBN 9781108041850]

Gwynne-Vaughan, Helen: *Fungi* (1922) [ISBN 9781108013215]

Henslow, John Stevens: *A Catalogue of British Plants Arranged According to
the Natural System* (1829) [ISBN 9781108061728]

Henslow, John Stevens: *A Dictionary of Botanical Terms* (1856)
[ISBN 9781108001311]

Henslow, John Stevens: *Flora of Suffolk* (1860) [ISBN 9781108055673]

Henslow, John Stevens: *The Principles of Descriptive and Physiological Botany*
(1835) [ISBN 9781108001861]

Hogg, Robert: *The British Pomology* (1851) [ISBN 9781108039444]

Hooker, Joseph Dalton, and Thomson, Thomas: *Flora Indica* (1855)
[ISBN 9781108037495]

Hooker, Joseph Dalton: *Handbook of the New Zealand Flora* (2 vols., 1864–7) [ISBN 9781108030410]

Hooker, William Jackson: *Icones Plantarum* (10 vols., 1837–54) [ISBN 9781108039314]

Hooker, William Jackson: *Kew Gardens* (1858) [ISBN 9781108065450]

Jussieu, Adrien de, edited by J.H. Wilson: *The Elements of Botany* (1849) [ISBN 9781108037310]

Lindley, John: *Flora Medica* (1838) [ISBN 9781108038454]

Müller, Ferdinand von, edited by William Woolls: *Plants of New South Wales* (1885) [ISBN 9781108021050]

Oliver, Daniel: *First Book of Indian Botany* (1869) [ISBN 9781108055628]

Pearson, H.H.W., edited by A.C. Seward: *Gnetales* (1929) [ISBN 9781108013987]

Perring, Franklyn Hugh et al.: *A Flora of Cambridgeshire* (1964) [ISBN 9781108002400]

Sachs, Julius, edited and translated by Alfred Bennett, assisted by W.T. Thiselton Dyer: *A Text-Book of Botany* (1875) [ISBN 9781108038324]

Seward, A.C.: *Fossil Plants* (4 vols., 1898–1919) [ISBN 9781108015998]

Tansley, A.G.: *Types of British Vegetation* (1911) [ISBN 9781108045063]

Traill, Catherine Parr Strickland, illustrated by Agnes FitzGibbon Chamberlin: *Studies of Plant Life in Canada* (1885) [ISBN 9781108033756]

Tristram, Henry Baker: *The Fauna and Flora of Palestine* (1884) [ISBN 9781108042048]

Vogel, Theodore, edited by William Jackson Hooker: *Niger Flora* (1849) [ISBN 9781108030380]

West, G.S.: *Algae* (1916) [ISBN 9781108013222]

Woods, Joseph: *The Tourist's Flora* (1850) [ISBN 9781108062466]

For a complete list of titles in the Cambridge Library Collection please visit:
www.cambridge.org/features/CambridgeLibraryCollection/books.htm

A N

A C C O U N T

O F T H E

F O X G L O V E,

A N D S O M E O F

Its Medical Uſes, &c.

a 2

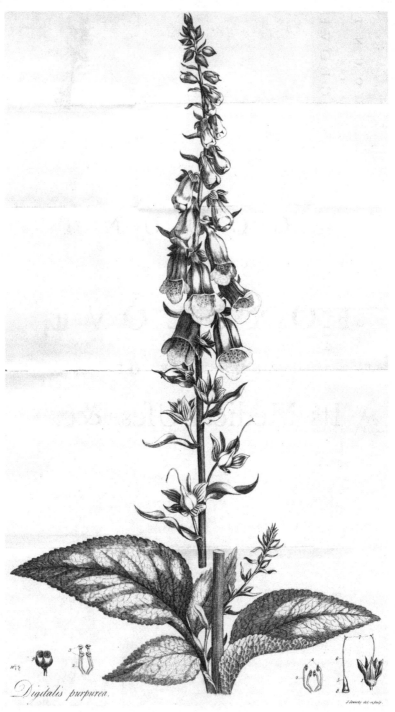

Digitalis purpurea.

The material originally positioned here is too large for reproduction in this reissue. A PDF can be downloaded from the web address given on page iv of this book, by clicking on 'Resources Available'.

A N

A C C O U N T

OF THE

F O X G L O V E,

A N D

Some of its Medical Uſes :

WITH

PRACTICAL REMARKS ON DROPSY,

AND OTHER DISEASES.

B Y

WILLIAM WITHERING, M. D.
Phyſician to the General Hoſpital at Birmingham.

—— *nonumque prematur in annum.*

HORACE.

BIRMINGHAM: PRINTED BY M. SWINNEY;
FOR
G. G. J. AND J. ROBINSON, PATERNOSTER-ROW, LONDON.
M,DCC,LXXXV.

PREFACE.

AFTER being frequently urged to write upon this fubject, and as often declining to do it, from apprehenfion of my own inability, I am at length compelled to take up the pen, however unqualified I may ftill feel myfelf for the tafk.

The ufe of the Foxglove is getting abroad, and it is better the world fhould derive fome inftruction, however imperfect, from my experience, than that the lives of men fhould be hazarded by its unguarded exhibition, or that a medicine of fo much efficacy fhould be condemned and rejected as dangerous and unmanageable.

It

It is now about ten years fince I firft began to ufe this medicine. Experience and
cautious attention gradually taught me how
to ufe it. For the laft two years I have not
had occafion to alter the modes of management ; but I am ftill far from thinking
them perfect.

It would have been an eafy tafk to have
given felect cafes, whofe fuccefsful treatment
would have fpoken ftrongly in favour of the
medicine, and perhaps been flattering to my
own reputation. But Truth and Science
would condemn the procedure. I have
therefore mentioned every cafe in which I
have prefcribed the Foxglove, proper or improper, fuccefsful or otherwife. Such a
conduct will lay me open to the cenfure of
thofe who are difpofed to cenfure, but it
will meet the approbation of others, who are
the beft qualified to be judges.

To the Surgeons and Apothecaries, with
whom I am connected in practice, both in
this town and at a diftance, I beg leave to
make

make this public acknowledgment, for the
affiftance they fo readily afforded me, in per-
fecting fome of the cafes, and in communicating the events of others.

The ages of the patients are not always
exact, nor would the labour of making them
fo have been repaid by any ufeful confequences. In a few inftances accuracy in that
refpect was neceffary, and there it has been
attempted ; but in general, an approxima-
tion towards the truth, was fuppofed to be
fufficient.

The cafes related from my own experience, are generally written in the fhorteft
form I could contrive, in order to fave time
and labour. Some of them are given more
in detail, when particular circumftances
made fuch detail neceffary ; but the cafes
communicated by other practitioners, are
given in their own words.

I muft caution the reader, who is not a
practitioner in phyfic, that no general de-
ductions, decifive upon the failure or fuccefs

b   of

of the medicine, can be drawn from the cafes I now prefent to him. Thefe cafes muft be confidered as the moft hopelefs and deplorable that exift ; for phyficians are feldom confulted in chronic difeafes, till the ufual remedies have failed : and, indeed, for fome years, whilft I was lefs expert in the management of the Digitalis, I feldom prefcribed it, but when the failure of every other method compelled me to do it ; fo that upon the whole, the inftances I am going to adduce, may truly be confidered as cafes loft to the common run of practice, and only fnatched from deftruction, by the efficacy of the Digitalis; and this in fo remarkable a manner, that, if the properties of that plant had not been difcovered, by far the greateft part 'of thefe patients muft have died.

There are men who will hardly admit of any thing which an author advances in fupport of a favorite medicine, and I allow they may have fome caufe for their hefitation; nor do I expect they will wave their ufual modes of

judg-

judging upon the prefent occafion. I could wifh therefore that fuch readers would pafs over what I have faid, and attend only to the communications from correfpondents, becaufe they cannot be fuppofed to poffefs any unjuft predilection in favour of the medicine : but I cannot advife them to this ftep, for I am certain they would then clofe the book, with much higher notions of the efficacy of the plant than what they would have learnt from me. Not that I want faith in the difcernment or in the veracity of my correfpondents, for they are men of eftablifhed reputation; but the cafes they have fent me are, with fome exceptions, too much felected. They are not upon this account lefs valuable in themfelves, but they are not the proper premifes from which to draw permanent conclufions.

I wifh the reader to keep in view, that it is not my intention merely to introduce a new diuretic to his acquaintance, but one which, though not infallible, I believe to be much more certain than any other in prefent ufe.

After

After all, in ſpite of opinion, prejudice, or error, Time will fix the real value upon this diſcovery, and determine whether I have impoſed upon myſelf and others, or contributed to the benefit of ſcience and mankind.

Birmingham, 1ſt *July*,
1785.

INTRO-

INTRODUCTION.

THE Foxglove is a plant fufficiently common in this ifland, and as we have but one fpecies, and that fo generally known, I fhould have thought it fuperfluous either to figure or defcribe it; had I not more than once feen the leaves of Mullein* gathered for thofe of Foxglove. On the continent of Europe too, other fpecies are found, and I have been informed that our fpecies is very rare in fome parts of Germany, exifting only by means of cultivation, in gardens.

Our plant is the *Digitalis purpurea* † of Linnæus. It belongs to the 2d order of the 14th clafs, or the DIDYNAMIA ANGIOSPERMIA. The *effential characters* of the genus are, *Cup with 5 divifions. Bloffom bell-fhaped, bulging. Capfule egg-fhaped, 2-celled.*— LINN.

DIGITALIS *purpu'rea*. Little leaves of the empalement egg-fhaped, fharp. Bloffoms blunt; the upper lip entire. LINN.

REFE-

---

b 3

REFERENCES TO FIGURES. Thefe are difpofed in the order of comparative excellence.

Riuini monopet. 104.
Flora danica, 74, *parts of fructification.*
Tournefort Inftitutiones. 73, *A*, *E*, *L*, *M.*
Fuchfii Hift. Plant. 893, *copied in*
Tragi ftirp. hiftor. 889.
J. Bauhini hiftor. Vol. ii. 812. 3, *and*
Lonicera 74, 1.
Blackwell. auct. 16.
Dodonæi pempt. ftirp. hift. 169, *reprinted in*
Gerard emacul. 790, 1, *and copied in*
Parkinfon Theatr. botanic. 653, 1.
Gerard, firft edition, 646, 1.
Hiftor. Oxon. Morifon. V. 8, *row* 1. 1.
Flor. danic. 74, *the reduced figure.*

Bloffom. The bellying part on the infide fprinkled with fpots like little eyes. *Leaves* wrinkled. LINN.

BLOSSOM. Rather tubular than bell-fhaped, bulging on the under fide, purple; the narrow tubular part at the bafe, white. *Upper lip* fometimes flightly cloven.

CHIVES. *Threads* crooked, white. *Tips* yellow.

POINTAL, *Seed-bud* greenifh. *Honey-cup* at its bafe more yellow. *Summit* cloven.

S. VESS. *Capfule* not quite fo long as the cup.

ROOT, Knotty and fibrous.

STEM.

STEM. About 4 feet high; obfcurely angular; leafy.

LEAVES. Slightly but irregularly ferrated, wrinkled; dark green above, paler underneath. *Lower leaves* egg-fhaped; upper leaves fpear-fhaped. *Leaf-ftalks* flefhy; bordered.

FLOWERS. Numerous, moftly growing from one fide of the ftem and hanging down one over another. *Floral-leaves* fitting, taper-pointed. The numerous purple bloffoms hanging down, mottled within; as wide and nearly half as long as the finger of a common-fized glove, are fufficient marks whereby the moft ignorant may diftinguifh this from every other Britifh plant; and the leaves ought not to be gathered for ufe but when the plant is in bloffom.

PLACE. Dry, gravelly or fandy foils; particularly on floping ground. It is a biennial, and flowers from the middle of *June* to the end of *July*.

I have not obferved that any of our cattle eat it. The root, the ftem, the leaves, and the flowers have a bitter herbaceous tafte, but I don't perceive that naufeous bitter which has been attributed to it.

This plant ranks amongft the LURIDÆ, one of the Linnæan orders in a natural fyftem. It has for congenera, NICOTIANA, ATROPA, HYOSCYAMUS, DATURA, SOLANUM, &c. fo that from the knowledge we poffefs of the virtues of thofe plants, and reafoning from botanical analogy, we might be led to guefs at fomething of its properties.

I in-

I intended in this place to have traced the history of its effects in difeafes from the time of Fuchfius, who firft defcribes it, but I have been anticipated in this intention by my very valuable friend, Dr. Stokes of Stourbridge, who has lately fent me the following

HISTORICAL VIEW of the Properties of Digitalis.

FUCHSIUS in his *hift. ftirp.* 1542, is the firft author who notices it. From him it receives its name of DIGITALIS, in allufion to the German name of *Fingerhut*, which fignifies a finger-ftall, from the bloffoms refembling the finger of a glove.

SENSIBLE QUALITIES. Leaves bitterifh, very naufeous. LEWIS *Mat. med.* i. 342.

SENSIBLE EFFECTS. Some perfons, foon after eating of a kind of omalade, into which the leaves of this, with thofe of feveral other plants, had entered as an ingredient, found themfelves much indifpofed and were prefently after attacked with vomitings, DODONÆUS *pempt.* 170.

It is a medicine which is proper only for ftrong conftitutions, as it purges very violently, and excites exceffive vomitings. RAY, *hift.* 767.

BOERHAAVE judges it to be of a poifonous nature, *hift. plant.* but Dr. ALSTON ranks it among thofe indigenous vegetables, " which, though now difre-
" garded,

" garded, are medicines of great virtue, and fcarce-
" ly inferior to any that the Indies afford." Lewis
Mat. med. i. *p.* 343.

Six or feven fpoonfuls of the decoction produce
naufea and vomiting, and purge ; not without
fome marks of a deleterious quality. Haller *hift. n.*
330 from *Aerial Infl. p.* 49, 50.

The following is an abridged Account of its Effects upon Turkeys.

M. Salerne, a phyfician at Orleans, having heard
that feveral turkey pouts had been killed by being
fed with Foxglove leaves, inftead of mullein, he
gave fome of the fame leaves to a large vigorous
turkey. The bird was fo much affected that he
could not ftand upon his legs, he appeared drunk,
and his excrements became reddifh. Good nou-
rifhment reftored him to health in eight days.

Being then determined to pufh the experiment
further, he chopped fome more leaves, mixed them
with bran, and gave them to a vigorous turkey cock
which weighed feven pounds. This bird foon appeared drooping and melancholy; his feathers ftared,
his neck became pale and retracted, The leaves
were given him for four days, during which time
he took about half a handful. Thefe leaves had
been gathered about eight days, and the winter was
far advanced. The excrements, which are naturally

ally green and well formed, became, from the firſt, liquid and reddiſh, like thoſe of a dyſenteric patient.

The animal refuſing to eat any more of this mix- ture which had done him ſo much miſchief, I was obliged to feed him with bran and water only; but notwithſtanding this, he continued drooping, and without appetite. At times he was ſeized with con- vulſions, ſo ſtrong as to throw him down; in the intervals he walked as if drunk; he did not attempt to perch, he uttered plaintive cries. At length he refuſed all nouriſhment. On the fifth or ſixth day the excrements became as white as chalk; after- terwards yellow, greeniſh, and black. On the eigh- teenth day he died, greatly reduced in fleſh, for he now weighed only three pounds.

On opening him we found the heart, the lungs, the liver, and gall-bladder ſhrunk and dried up; the ſtomach was quite empty, but not deprived of its villous coat. *Hiſt. de l'Academ.* 1748. *p.* 84.

EPILEPSY.—" It hath beene of later experience
" found alſo to be effectual againſt the falling ſick-
" neſſe, that divers have been cured thereby; for
" after the taking of the *Decoct. manipulor, ii. c. poly-*
" *pod. quercin. contus.* ℥*iv. in cercviſia,* they that have
" been troubled with it twenty-ſix years, and have
" fallen once in a weeke, or two or three times in a
" moneth, have not fallen once in fourteen or fif-
" teen moneths, that is until the writing hereof."
Parkinſon, p. 654.
SCROPHULA.—

SCROPHULA.—" The herb bruifed, or the juice
" made up into an ointment, and applied to the
" place, hath been found by late experience to be
" availeable for the King's Evill," PARK. p. 654.

Several hereditary inftances of this difeafe faid
to have been cured by it.   AEREAL INFLUENCES, *p*
49, 50, quoted by HALLER, *hiſt. n.* 330.

A man with *fcrophulous ulcers* in various parts of
the body, and which in the right leg were fo viru-
lent that its amputation was propofed, cured by.
*ſucc. exprefs. cochl. i. bis intra xiv. dies, in ½ pinta
cerevifiæ calidæ.*

The leaves remaining after the preffing out of the
juice, were applied every day to the ulcers. *Pract.
eſs. p.* 40. quoted by MURRAY *apparat. medicam. i. p.*
491.

A young woman with a *fcrophulous tumour of the
eye,* a remarkable *fwelling of the upper lip, and painful
tumours of the joints of the fingers,* much relieved ;
but the medicine was left off, on account of its violent effects on the conftitution. *Ib. p.* 42 quoted as
above.

A man with a *fcrophulous tumour of the right elbow,*
attended for three years *with excruciating pains,* was
nearly cured by four dofes of the juice taken once
a month. *Ib. p.* 43. as above.

The phyficians and furgeons of the Worcefter Infirmary have employed it in ointments and poul-
tices with remarkable efficacy. *Ib. p.* 44. It was re-
com-

commended to them by Dr. Baylies of Evefham, now of Berlin, as a remedy for this difeafe. Dr. Wall gave it a tryal as well externally as internally, but their experiments did not lead them to obferve any other properties in it, than thofe of a highly naufeating medicine and draftic purgative.

WOUNDS. In confiderable eftimation for the healing all kinds of wounds, *Lobel. adv.* 245.

Principally of ufe in ulcers, which difcharge confiderably, being of little advantage in fuch as are dry. HULSE, in R. hift. 768.

DOCTOR BAYLIES, phyfician to his Pruffian Majefty, informed me, when at Berlin, that he employed it with great fuccefs in caries, and obftinate fore legs.

DYSPNŒA *Pituitofa* Sauvages i. 657.—" Boiled " in water, or wine, and drunken doth cut and " confume the thicke toughneffe of groffe, and " flimie flegme, and naughtie humours. The " fame, or boiled with honied water or fugar, doth " fcoure and clenfe the breft, ripeneth and bring- " eth foorth tough and clammie flegme. It open- " eth alfo the ftoppage of the liver fpleene and " milt, and of the inwarde parts," GERARDE hift. " ed. I. p. 647.

" Whenfoever there is need of a rarefying or " extenuating of tough flegme or vifcous humours " troubling the cheft,—the decoction or juice here- " of made up with fugar or honey is availeable, as " alfo to clenfe and purge the body both upwards
                                                    " and

" and downwards fometimes, of tough flegme, and
" clammy humours, notwithftanding that thefe
" qualities are found to bee in it, there are but few
" phyfitions in our times that put it to thefe ufes,
" but it is in a manner wholly negleded."

PARKINSON, p. 654.

Previous to the year 1777, you informed me of
the great fuccefs you had met with in curing dropfies by means of the fol. Digitalis, which you then
confidered as a more certain diuretic than any you
had ever tried. Some time afterwards, Mr. Ruffel,
furgeon, of Worcefter, having heard of the fuccefs which had attended fome cafes in which you
had given it, requefted me to obtain for him any
information you might be inclined to communicate
refpecting its ufe. In confequence of this application, you wrote to me in the following terms.*

In a letter which I received from you in London,
dated *September* 29, 1778, you write as follows:—
" I wifh it was as eafy to write upon the Digitalis—
" I defpair of pleafing myfelf or inftructing others,
" in a fubject fo difficult. It is much eafier to
" write upon a difeafe than upon a remedy. The
" former is in the hands of nature, and a faithful
" obferver, with an eye of tolerable judgment,
" cannot fail to delineate a likenefs. The latter
" will ever be fubject to the whims, the inaccura-
" cies, and the blunders of mankind."—

In

* See the extract from this letter at page 5.

In my notes I find the following memorandum—
" *February* 20th, 1779, gave an account of Doctor
" Withering's practice, with the precautions ne-
" ceffary to its fuccefs, to the Medical Society at
" Edinburgh."—In the courfe of that year, the Di-
gitalis was prefcribed in the Edinburgh Infirmary, by
Dr. Hope, and in the following year, whilft I was
Clerk to Dr. Home, as Clinical Profeffor, I had a
favourable opportunity of obferving its fenfible ef-
fects.

In one cafe in which it was given properly at firft,
the urine began to flow freely on the fecond day.
On the third, the fwellings began to fubfide. The
dofe was then increafed more than *quadruple* in the
twenty-four hours. On the fifth day ficknefs came
on, and much purging, but the urine ftill increafed
though the pulfe funk to 50. On the 7th day, a
quadruple dofe of the infufion was ordered to be taken
every third hour, fo as to bring on naufea again.
The pulfe fell to forty-four, and at length to thirty-
five in a minute. The patient gradually funk and
died on the fixteenth day ; but previous to her
death, for two or three days, her pulfe rofe to near
one hundred.—It is needlefs to obferve to you, how
widely the treatment of this cafe differed from the
method which you have found fo fuccefsful.

OF

OF THE PLATE.

THE figure of the Foxglove, facing the Title
Page, is copied by the permiffion and under
the infpection of Mr. Curtis, from his admirable
work, entitled FLORA LONDINENSIS. The accuracy
of the drawings, the beauty of the colouring, the full
defcriptions, the accurate fpecific diftinctions, and
the ufes of the different plants, cannot fail to recom-
mend that work to the patronage of all who are in-
terefted in the encouragement of genius, or the
promotion of ufeful knowledge.

EXPLANATION.

Fig. 1. The Empalement.

Fig. 2, 3, 4. Four CHIVES two long and two fhort,
TIPS at firft large, turgid, oval, touching at
bottom, of a yellowifh colour, and often fpot-
ted; laftly changing both their form and fituation in a fingular manner.

Fig. 5, 6, 7. SEED-BUD rather conical, of a yellow
green colour. *Shaft* fimple. *Summit* cloven,

Fig. 8. *Honeycup* a gland, furrounding the bottom
of the Seed-bud.

Fig. 9. SEED-VESSEL, a pointed oval *Capfule*, of two
cells and two valves, the lowermoft valve fplit-
ting in two.

Fig. 10. SEEDS numerous, blackifh, fmall, lopped
at each end.

AN

ACCOUNT

OF THE

INTRODUCTION of FOXGLOVE

INTO

MODERN PRACTICE.

AS the more obvious and fenfible properties of plants, fuch as colour, tafte, and fmell, have but little connexion with the difeafes they are adapted to cure; fo their peculiar qualities have no certain dependence upon their external configuration. Their chemical examination by fire, after an immenfe wafte of time and labour, having been found ufelefs, is now abandoned by general confent. Poffibly other modes of analyfis will be found out, which may turn to better account; but we have hitherto made only a very fmall progrefs in the chemiftry of animal and vegetable fubftances. Their virtues muft therefore be learnt, either from obferving their effects upon infects and quadrupeds; from analogy, deduced from the already known powers of fome of their congenera, or from the empirical ufages and experience of the populace.

The firft method has not yet been much attended to; and the fecond can only be perfected in proportion as we approach towards the difcovery of a truly natural fyftem; but the laft, as far as it extends, lies

A within

within the reach of every one who is open to information, regardlefs of the fource from whence it
fprings.

It was a circumftance of this kind which firft fixed
my attention on the Foxglove.

In the year 1775, my opinion was afked concern-
ing a family receipt for the cure of the dropfy. I
was told that it had long been kept a fecret by an
old woman in Shropfhire, who had fometimes made
cures after the more regular practitioners had failed.
I was informed alfo that the effects produced were
violent vomiting and purging; for the diuretic effects feemed to have been overlooked. This medicine was compofed of twenty or more different herbs;
but it was not very difficult for one converfant in
thefe fubjects, to perceive, that the active herb could
be no other than the Foxglove.

My worthy predeceffor in this place, the very hu-
mane and ingenious Dr. Small, had made it a practice to give his advice to the poor during one hour
in a day. This practice, which I continued until
we had an Hofpital opened for the reception of the
fick poor, gave me an opportunity of putting my
ideas into execution in a variety of cafes; for the
number of poor who thus applied for advice,
amounted to between two and three thoufand an-
nually. I foon found the Foxglove to be a very
powerful diuretic; but then, and for a confiderable
time afterwards, I gave it in dofes very much too
large

large, and urged its continuance too long; for mif-
led by reafoning from the effects of the fquill, which
generally acts beft upon the kidneys when it excites
naufea, I wifhed to produce the fame effect by the
Foxglove. In this mode of prefcribing, when I had
fo many patients to attend to in the fpace of one,
or at moft of two hours, it will not be expected that
I could be very particular, much lefs could I take
notes of all the cafes which occurred. Two or three
of them only, in which the medicine fucceeded, I
find mentioned amongft my papers. It was from
this kind of experience that I ventured to affert, in
the Botanical Arrangement publifhed in the courfe of
the following fpring, that the Digitalis purpurea
" merited more attention than modern practice be-
" ftowed upon it."

I had not, however, yet introduced it into the more
regular mode of prefcription; but a circumftance
happened which accelerated that event. My truly
valuable and refpectable friend, Dr. Afh, informed
me that Dr. Cawley, then principal of Brazen Nofe
College, Oxford, had been cured of a Hydrops Pectoris, by an empirical exhibition of the root of the
Foxglove, after fome of the firft phyficians of the age
had declared they could do no more for him. I was
now determined to purfue my former ideas more
vigoroufly than before, but was too well aware of
the uncertainty which muft attend on the exhibition
of the *root* of a *biennial* plant, and therefore continued to ufe the *leaves*. Thefe I had found to vary
much as to dofe, at different feafons of the year;

but I expected, if gathered always in one condition
of the plant, viz. when it was in its flowering state,
and carefully dried, that the dose might be ascer-
tained as exactly as that of any other medicine; nor
have I been disappointed in this expectation. The
more I saw of the great powers of this plant, the
more it seemed necessary to bring the doses of it to
the greatest possible accuracy. I suspected that this
degree of accuracy was not reconcileable with the
use of a *decoction*, as it depended not only upon the
care of those who had the preparation of it, but it
was easy to conceive from the analogy of another
plant of the same natural order, the tobacco, that
its active properties might be impaired by long boiling. The decoction was therefore discarded, and
the *infusion* substituted in its place. After this I began to use the leaves in *powder*, but I still very often
prescribe the infusion.

Further experience convinced me, that the *diuretic* effects of this medicine do not at all depend upon its exciting a nausea or vomiting; but, on the
contrary, that though the increased secretion of
urine will frequently succeed to, or exist along with
these circumstances, yet they are so far from being
friendly or necessary, that I have often known the
discharge of urine checked, when the doses have
been imprudently urged so as to occasion sickness.

If the medicine purges, it is almost certain to fail
in its desired effect; but this having been the case,
I have seen it afterwards succeed when joined with
small

fmall dofes of opium, fo as to reftrain its action on the bowels.

In the fummer of the year 1776, I ordered a quantity of the leaves to be dried, and as it then became poffible to afcertain its dofes, it was gradually adopted by the medical practitioners in the circle of my acquaintance.

In the month of *November* 1777, in confequence of an application from that very celebrated furgeon, Mr. Ruffel, of Worcefter, I fent him the following account, which I choofe to introduce here, as fhewing the ideas I then entertained of the medicine, and how much I was miftaken as to its real dofe.—
" I generally order it in decoction. Three drams of
" the dried leaves, collected at the time of the blof-
" foms expanding, boiled in twelve to eight ounces of
" water. Two fpoonfuls of this medicine, given eve-
" ry two hours, will fooner or later excite a naufea.
" I have fometimes ufed the green leaves gathered in
" winter, but then I order three times the weight;
" and in one inftance I ufed three ounces to a pint
" decoction, before the defired effect took place. I
" confider the Foxglove thus given, as the moft cer-
" tain diuretic I know, nor do its diuretic effects
" depend merely upon the naufea it produces, for
" in cafes where fquill and ipecac. have been fo
" given as to keep up a naufea feveral days together,
" and the flow of urine not taken place, I have found
" the Foxglove to fucceed; and I have, in more than
" one inftance, given the Foxglove in fmaller and
A 3   " more

" more diftant dofes, fo that the flow of urine has
" taken place without any fenfible affection of the
" ftomach; but in general I give it in the manner
" firft mentioned, and order one dofe to be taken
" after the ficknefs commences. I then omit all me-
" dicines, except thofe of the cordial kind are wanted,
" during the fpace of three, four, or five days. By
" this time the naufea abates, and the appetite be-
" comes better than it was before. Sometimes the
" brain is confiderably affected by the medicine, and
" indiftinct vifion enfues; but I have never yet
" found any permanent bad effects from it."—

" I ufe it in the Afcites, Anafarca, and Hydrops
" Pectoris; and fo far as the removal of the water
" will contribute to cure the patient, fo far may be
" expected from this medicine: but I wifh it not to
" be tried in afcites of female patients, believing
" that many of thefe cafes are dropfies of the ovaria;
" and no fenfible man will ever expect to fee thefe
" encyfted fluids removed by any medicine."

" I have often been obliged to evacuate the water
" repeatedly in the fame patient, by repeating the
" decoction; but then this has been at fuch diftances
" of time as to allow of the interference of other
" medicines and a proper regimen, fo that the patient
" obtains in the end a perfect cure. In thefe cafes
" the decoction becomes at length fo very difagree-
" able, that a much fmaller quantity will produce the
" effect, and I often find it neceffary to alter its
" tafte by the addition of Aq. Cinnam. fp. or Aq.
" Juniper. compofita."   " I al-

" I allow, and indeed enjoin my patients to drink
" very plentifully of fmall liquors through the whole
" courfe of the cure; and fometimes, where the eva-
" cuations have been very fudden, I have found a
" bandage as neceffary as in the ufe of the trochar."—

Early in the year 1779, a number of dropfical
cafes offered themfelves to my attention, the confequences of the fcarlet fever and fore throat which
had raged fo very generally amongft us in the preceding year.   Some of thefe had been cured by
fquills or other diuretics, and relapfed; in others,
the dropfy did not appear for feveral weeks after the
original difeafe had ceafed: but I am not able to
mention many particulars, having omitted to make
notes.   This, however, is the lefs to be regretted,
as the fymptoms in all were very much alike, and
they were all without an exception cured by the Foxglove.

This laft circumftance encouraged me to ufe the
medicine more frequently than I had done heretofore, and the increafe of practice had taught me to
improve the management of it.

In *February* 1779, my friend, Dr. Stokes, communicated to the Medical Society at Edinburgh the re-
fult of my experience of the Foxglove; and, in a letter addreffed to me in *November* following, he fays,
" Dr. Hope, in confequence of my mentioning its
" ufe to my friend, Dr. Broughton, has tried the
" Foxglove in the Infirmary with fuccefs." Dr.
Stokes

Stokes alfo tells me that Dr. Hamilton cured Dropfies with it in the year 1781.

I am informed by my very worthy friend Dr. Duncan, that Dr. Hamilton, who learnt its ufe from Dr. Hope, has employed it very frequently in the Hofpital at Edinburgh. Dr. Duncan alfo tells me, that the late very ingenious and accomplifhed Mr. Charles Darwin, informed him of its being ufed by his father and myfelf, in cafes of Hydrothorax, and that he has ever fince mentioned it in his lectures. and fometimes employed it in his practice.

At length, in the year 1783, it appeared in the new edition of the Edinburgh Pharmacopœia, into which, I am told, it was received in confequence of the recommendation of Dr. Hope. But from which, I am fatisfied, it will be again very foon rejected, if it fhould continue to be exhibited in the unre-ftrained manner in which it has heretofore been ufed at Edinburgh, and in the enormous dofes in which it is now directed in London.

In the following cafes the reader will find other difeafes befides dropfies; particularly feveral cafes of confumption. I was induced to try it in thefe, from being told, that it was much ufed in the Weft of England, in the Phthifis Pulmonalis, by the common people. In this difeafe, however, in my hands, it has done but little fervice, and yet I am difpofed to wifh it a further trial, for in a copy of Parkinfon's Herbal, which I faw about two years ago, I found

I found the following manufcript note at the article Digitalis, written, I believe, by a Mr. Saunders, who practifed for many years with great reputation as a furgeon and apothecary at Stourbridge, in Worcefterfhire.

" Confumptions are cured infallibly by weak de-
" coction of Foxglove leaves in water, or wine and
" water, and drank for conftant drink. Or take of
" the juice of the herb and flowers, clarify it, and
" make a fine fyrup with honey, of which take
" three fpoonfuls thrice in a day, at phyfical hours.
" The ufe of thefe two things of late has done, in
" confumptive cafes, great wonders. But be cautious
" of its ufe, for it is of a vomiting nature. In
" thefe things begin fparingly, and increafe the dofe
" as the patient's ftrength will bear, leaft, inftead of
" a fovereign medicine, you do real damage by this
" infufion or fyrup."

The precautions annexed to his encomiums of this medicine, lead one to think that he has fpoken from his own proper experience.

I have lately been told, that a perfon in the neighbourhood of Warwick, poffeffes a famous family receipt for the dropfy, in which the Foxglove is the active medicine; and a lady from the weftern part of Yorkfhire affures me, that the people in her country often cure themfelves of dropfical complaints by drinking Foxglove tea. In confirmation of this, I recollect about two years ago being defired to vifit a
travelling

travelling Yorkſhire tradeſman. I found him inceſ-
fantly vomiting, his viſion indiſtinct, his pulſe forty
in a minute. Upon enquiry it came out, that his
wife had ſtewed a large handful of green Foxglove
leaves in half a pint of water, and given him the
liquor, which he drank at one draught, in order to
cure him of an aſthmatic affection. This good woman knew the medicine of her country, but not
the doſe of it, for her huſband narrowly eſcaped
with his life.

It is probable that this rude mode of exhibiting
the Foxglove has been more general than I am at
preſent aware of; but it is wonderful that no author
ſeems to have been acquainted with its effects as a
diuretic.

C A S E S,

C A S E S,

In which the Digitalis was given by the Direction of the Author.

1 7 7 5.

I T was in the courfe of this year that I began to ufe the Digitalis in dropfical cafes. The patients were fuch as applied at my houfe for advice gratis. I cannot pretend to charge my memory with particular cafes, or particular effects, and I had not leifure to make notes. Upon the whole, however, it may be concluded, that the medicine was found ufeful, or I fhould not have continued to employ it.

C A S E I.

December 8th. A man about fifty years of age, who had formerly been a builder, but was now much reduced in his circumftances, complained to me of an afthma which firft attacked him about the latter end of autumn. His breath was very fhort, his countenance was funken, his belly large; and, upon examination, a fluctuation in it was very perceptible. His urine for fome time paft had been fmall in quantity. I directed a decoction of Fol. Digital. recent. which made him very fick, the ficknefs recurring at intervals for feveral days, during which time he made a large quantity of water. His breath gradually drew eafier, his belly fubfided, and in
about

about ten days he began to eat with a keen appetite. He afterwards took fteel and bitters.

1776.
C A S E   II.

January 14th. A poor man labouring under an afcites and anafarca, was directed to take a decoction of Digitalis every four hours. It purged him fmartly, but did not relieve him. An opiate was now ordered with each dofe of the medicine, which then acted upon the kidneys very freely, and he foon loft all his complaints.

C A S E   III.

March 15th. A poor boy, about nine years of age, was brought for my advice. His countenance was pale, his pulfe quick and feeble, his body greatly emaciated, except his belly, which was very large, and, upon examination, contained a fluid. The cafe had been confidered as arifing from worms. He was directed to take the decoction of Digitalis night and morning. It operated as a diuretic, never made him fick, and he got well without any other medicine.

C A S E   IV.

July 25th. Mrs. H———, of A———, near N———, between forty and fifty years of age, a few weeks ago, after fome previous indifpofition, was attacked by a fevere cold fhivering fit, fucceeded by fever; great pain in her left fide, fhortnefs of breath, perpetual cough, and, after fome days,

copious

copious expectoration. On the 4th of *June*, Dr. Darwin,* was called to her. I have not heard what was then done for her, but, between the 15th of *June*, and 25th of *July*, the Doctor, at his different visits, gave her various medicines of the deobstruent, tonic, antispasmodic, diuretic, and evacuant kinds.

On the 25th of *July* I was desired to meet Dr. Darwin at the lady's house. I found her nearly in a state of suffocation; her pulse extremely weak and irregular, her breath very short and laborious, her countenance sunk, her arms of a leaden colour, clammy and cold. She could not lye down in bed, and had neither strength nor appetite, but was extremely thirsty. Her stomach, legs, and thighs were greatly swollen; her urine very small in quantity, not more than a spoonful at a time, and that very seldom. It had been proposed to scarify her legs, but the proposition was not acceded to.

She had experienced no relief from any means that had been used, except from ipecacoanha vomits; the dose of which had been gradually increased from 15 to 40 grains, but such was the insensible state of her stomach for the last few days, that even those very large doses failed to make her sick, and consequently purged her. In this situation of things I knew of nothing likely to avail us, except the Digitalis: but this I hesitated to propose, from an apprehension that little could be expected from any thing; that an unfavourable termination would tend to discredit

---

* Then resident at Lichfield, now at Derby.

difcredit a medicine which promifed to be of great
benefit to mankind, and I might be cenfured for a
prefcription which could not be countenanced by
the experience of any other regular practitioner.
But thefe confiderations foon gave way to the defire
of preferving the life of this valuable woman, and
accordingly I propofed the Digitalis to be tried;
adding, that I fometimes had found it to fucceed
when other, even the moft judicious methods, had
failed.   Dr. Darwin very politely, acceded immediately to my propofition, and, as he had never
feen it given, left the preparation and the dofe to
my direction.   We therefore prefcribed as follows:

R. Fol. Digital. purp. recent. ʒiv. coque ex
    Aq. fontan. puræ ℔ifs ad ℔i. et cola.
R. Decoct. Digital. ʒifs.
    Aq. Nuc. Mofchat. ʒii. M. fiat. hauft. 2dis
horis fumend.

The patient took five of thefe draughts, which
made her very fick, and acted very powerfully upon the kidneys, for within the firft twenty-four
hours fhe made upwards of eight quarts of water.
The fenfe of fulnefs and oppreffion acrofs her ftomach was greatly diminifhed, her breath was eafed,
her pulfe became more full and more regular, and
the fwellings of her legs fubfided.

26th.   Our patient being thus fnatched from im-
pending deftruction, Dr. Darwin propofed to give
her a decoction of pareira brava and guiacum fhav-
ings,

ings, with pills of myrrh and white vitriol; and, if coftive, a pill with calomel and aloes. To thefe propofitions I gave a ready affent.

30th. This day Dr. Darwin faw her, and directed a continuation of the medicines laft prefcribed.

Augufl 1ft. I found the patient perfectly free from every appearance of dropfy, her breath quite eafy, her appetite much improved, but ftill very weak. Having fome fufpicion of a difeafed liver, I directed pills of foap, rhubarb, tartar of vitriol, and calomel to be taken twice a day, with a neutral faline draught.

9th. We vifited our patient together, and repeated the draughts directed on the 26th of *June*, with the addition of tincture of bark, and alfo ordered pills of aloes, guiacum, and fal martis to be taken if coftive.

September 10th. From this time the management of the cafe fell entirely under my direction, and perceiving fymptoms of effufion going forwards, I defired that a folution of merc. fubl. corr. might be given twice a day.

19th. The increafe of the dropfical fymptoms now made it neceffary to repeat the Digitalis. The dried leaves were ufed in infufion, and the water was prefently evacuated, as before.

It

It is now almoft nine years fince the Digitalis was firft prefcribed for this lady, and notwithftanding I have tried every preventive method I could devife, the dropfy ftill continues to recur at times ; but is never allowed to increafe fo as to caufe much dif-trefs, for fhe occafionally takes the infufion and re-lieves herfelf whenever fhe choofes. Since the firft exhibition of that medicine, very fmall dofes have been always found fufficient to promote the flow of urine.

I have been more particular in the narrative of this cafe, partly becaufe Dr. Darwin has related it ra-ther imperfectly in the notes to his fon s pofthumous publication, trufting, I imagine, to memory, and partly becaufe it was a cafe which gave rife to a ve-ry general ufe of the medicine in that part of Shrop-fhire.

C A S E   V.

December 10th. Mr. L———, Æt. 35. Afcites and anafarca, the confequence of very intemperate living. After trying fquill and other medicines to no purpofe, I directed a decoction of the Fol. Digi-tal. recent. fix drams to a pint; an eighth part to be taken every fourth hour. This made him fick, and produced a copious flow of urine, but not enough to remove all the dropfical fymptoms. After a fort-night a ftronger decoction was ordered, and, upon a third trial, as the winter advanced, it became neceffary to ufe four ounces to the pint decoction; and thus he got free from all his complaints.

In

In *October* 1777, in confequence of having purfued his intemperate mode of living, his dropfy returned, accompanied by evident marks of difeafed
vifcera. A decoction of two drams of Fol. Digital.
ficcat. to a pint, once more removed the dropfy. He
took a wine glafs full thrice a day.

In *January* 1778, I was defired to vifit him again.
I found he had gone on in his ufual intemperate life,
his countenance jaundiced, and the dropfy coming
on apace. After giving fome deobftruent medicines, I again directed the Digitalis, which again
emptied the water; but he did not furvive many
weeks.

1777.

C A S E VI.

February —. Mrs. M———, Æt. 45. Afcites
and anafarca, but not much otherwife difeafed, and
well enough to walk about the houfe, and fee after
her family affairs. I thought this a fair cafe for a
trial of the Digitalis, and therefore directed a decoction of the frefh leaves, the ftock of dried ones
being exhaufted. About a week afterwards, calling
to fee my patient, I was informed that fhe was dead;
that the third day after my firft vifit fhe fuddenly
fell down, and expired. Upon enquiry I found
fhe had not taken any of the medicine; for the
fnow had lain fo deep upon the ground, that the
apothecary had not been able to procure it. Had

B the

the medicine been given in a cafe feemingly fo favourable as this, and had the patient died under its
ufe, is it not probable that the death would have
been attributed to it?

C A S E   VII.

February 11th. Mr. E——, of W——, Æt. 61.
Hydrothorax, afcites and anafarca, confequences of
hard drinking. He had been attended for fome
time by a phyfician in his neighbourhood, who had
treated his cafe with the ufual remedies, but without affording him any relief; nor could I expect to
fucceed better by any other medicine than the Digitalis. The dried leaves were not to be had; and
the green ones at this feafon being very uncertain in
their ftrength, I ordered four ounces of the roots
in a pint decoction, and directed three fpoonfuls to
be given every fourth hour, until it either excited
naufea, or a free difcharge of urine; both thefe
effects took place nearly at the fame time: he made
a large quantity of water, the fwellings fubfided
very confiderably, and his breath became eafy. Eight
days afterwards he began upon a courfe of bitters
and deobftruents. The dropfical fymptoms foon
increafed again, but he had fuffered fo much from
the feverity of the ficknefs before, that he was neither willing to take, nor I to give the fame medicine
again.

Perhaps this patient might have been faved, if I
had been well acquainted with the management and
real

real dofes of the medicine, which was certainly in this inftance made very much too ftrong; and notwithftanding the caution to ftop the further exhibition when certain effects fhould take place, it feems the quantity previoufly fwallowed was fufficient to diftrefs him exceedingly.

C A S E VIII.

March 11th. Mrs. H———, Æt. 32. A few days after a tedious labour, had her legs and thighs fwelled to a very great degree; pale and femi-tranf-parent,* with pain in both groins. After a purge of calomel and rhubarb, ung. merc. was ordered to be rubbed upon the groins, and the following decoction was directed:

R. Fol. Digital. purp. recent. ʒii.
    Aq. puræ. ℔i. coque ad ℔ifs et colatur. adde.
    Aq. cinn. fp. ʒiv. M. capiat. cyath. vinos.
parv. bis quotidie.

The decoction prefently increafed the fecretion of urine, and abated the diftenfion of the legs: in a fortnight the fwelling was gone; but fome days after leaving her bed, her legs fwelled again about the ancles, which was removed by another bottle of the decoction on the 21ft of *April*.

---

* This difeafe has lately been well defcribed by Mr. White, of Manchefter.

C A S E   IX.

March 29th. Mr. G———, Æt. 47. Very much deformed; afthma of feveral years continuance, but now dropfical to a great degree. Took feveral medicines without relief, and then tried the Digitalis, but with no better fuccefs.

C A S E   X.

April 10th. G—G———, Æt. 70. Afthma and anafarca. Took a decoction of the frefh leaves of the Digitalis, which produced violent ficknefs, but no immediate evacuation of water. After the ficknefs had ceafed altogether, the urine began to flow copioufly, and he was cured.

C A S E   XI.

July 10th. Mr. M——— of T———, Æt. 54. A very hard drinker; had been affected fince *November* laft with afcites and anafarca, for which he had taken feveral medicines without benefit. A decoction of the recent leaves of the Digitalis was then directed, an ounce and half to a pint, one eighth of which I ordered to be given every fourth hour. A few dofes brought on great naufea, indiftinct vifion, and a great flow of urine, fo as prefently to empty him of all the dropfical water. Indeed the evacuation was fo rapid and fo complete, that it became neceffary to apply a bandage round the belly, and to fupport him with cordials.

In

In fomething more than a year and a half, his
dropfy returned, but the Digitalis did not then
fucceed to our wifhes.   In *Auguft*, 1779, he was
tapped, and lived afterwards only about five weeks.

For more particulars, fee the extract of a letter
from Mr. Lyon.

C A S E   XII.

September 12th.   Mifs C—— of T——, Æt. 48.
An ovarium dropfy, and anafarcous legs and thighs.
For three months in the beginning of this year fhe
had been under the care of Dr. Darwin, who at
different times had given her blue vitriol, elaterium,
and calomel; decoction of pareira brava, and guiacum wood, with tincture of cantharides ; oxymel of
fquills, decoction of parfley roots, &c.   Finding no
relief, fhe difcontinued the ufe of medicines, until
the urgency of her fymptoms induced her to afk
my advice about the end of *Auguft*.   She was greatly emaciated, and had almoft a total lofs of appetite.
I firft tried fmall dofes of Merc. fublim. corr. in
folution, with decoction of burdock roots, and blifters to the thighs.   No advantage attending the
ufe of this plan, I directed a decoction of Fol.
Digit. a dram and half to a pint; one ounce to be
taken twice a day.   It prefently reduced the anafarcous fwellings, but made no alteration in the diftenfion of the abdomen.

B 3   C A S E

C A S E   XIII.

October 9th. Mrs. B———, Æt. 40. An ovarium dropfy. Took a decoction of Digitalis without effect. Her life was preferved for fome years by repeated tapping.

1 7 7 8.

C A S E   XIV.

February 8th. Mr. R——— of K———. Had formerly fuffered much from gout, and lived very intemperately. Jaundiced countenance ; afcites ; legs and thighs greatly fwollen ; appetite none ; extremely weak ; confined to his bed. Had taken many medicines from his apothecary without advantage. I ordered him decoction of Digitalis, and a cordial ; but he furvived only a few days.

C A S E   XV.

March 13th. Mr. M———, Æt. 54. A thorax greatly deformed ; afthma through the winter, fucceeded by dropfy in belly and legs. Pulfe very fmall ; face leaden coloured ; cough almoft continual. Decoction of feneka was directed, and fmall dofes of Dover's powder at night.

17th. Gum-ammoniac and fquill, with elixir paregor. at night.—26th, Squill and decoction of feneka.—30th, His complaints ftill increafing, decoction

coction of Digitalis was then directed, which relieved him in a few days; but his complaints returned again, and he died in the month of *June*.

C A S E XVI.

August 18th. Mr. B——, Æt. 33. Pulmonary confumption and dropfy. The Digitalis, and that failing, other diuretics were ufed, in hopes of gaining fome relief from the diftrefs occafioned by the dropfical fymptoms; but none of them were effectual. He was then attended by another phyfician, and died in about two months.

C A S E XVII.

September 21ft. Mrs. M—— W—— G——, Æt. 50. An ovarium dropfy. She took half a pint of Infuf. Digitalis, which made her fick, but did not increafe the quantity of urine. She was afterwards relieved by tapping.

C A S E XVIII.

October 28th. R—— W——, Æt. 33. Afcites and univerfal anafarca; countenance quite pale and bloated; appetite none, and the little food he forces down is generally rejected.

R. Fol. Digit. purp. ficcat. ʒiii.
    Aq. bull. ℔i. digere per horas duas, et colat. adde aq. junip. comp. ʒiii.

He

He was directed to take one ounce of this infusion every two hours until it should make him sick. This was on Wednesday. The fifth dose made him vomit. On Thursday afternoon he vomited again very freely, without having taken any more of the medicine. On Friday and Saturday he made more water than he had done for a week before, and the swellings of his face and body were confiderably abated. He was directed to omit all medicine so long, as the urine continued to flow freely, and also to keep an account of the quantity he made in twenty-four hours.

Thefe were his reports:

October 31ft.	Saturday,	5 half pints.
November 1ft.	Sunday,	6
2d.	Monday,	8
3d.	Tuefday,	8
4th.	Wednefday,	7
5th.	Thurfday,	8

On Wednefday he began to purge, and the purging ftill continues, but his appetite is better than he has known it for a long time. No fwelling remains but about his ancles, extending at night half way up his legs.

Omit all medicines at prefent.

7th.	Saturday,	7½ half pints.
8th.	Sunday,	8
9th.	Monday,	6¾
10th.	Tuefday,	6½
11th.	Wednefday,	6
12th.	Thurfday,	6¼

On

On Tuefday the 17th, fome fwelling ftill remained about his ancles, but he was in every other refpect perfectly well.

He took a few more dofes of the infufion, and no other medicine.

C A S E   XIX.

December 8th. W——— B———, Æt. 60. A hard drinker. Difeafed vifcera; afcites and anafarca. An infufion of Digitalis was directed, but it had no other effect than to make him fick.

1779.

In the beginning of this year we had many dropfies in children, who had fuffered from the Scarlatina Anginofa; they all yielded very readily to the Digitalis, but in fome the medicine purged, and then it did not prove diuretic, nor did it remove the dropfy until opium was joined with it, fo as to prevent it purging. ———I did not keep notes of thefe cafes, but I do not recollect a fingle inftance in which the Digitalis failed to effect a cure.

C A S E   XX.

January 1ft. Mr. H———. Hydrops Pectoris; legs and thighs prodigioufly anafarcous; a very diftreffing fenfe of fulnefs and tightnefs acrofs his ftomach; urine in fmall quantity; pulfe intermitting; breath very fhort.

He

He had taken various medicines, and been blif-
tered, but without relief. His complaints continuing to increafe, I directed an infufion of Digitalis,
which made him very fick; acted powerfully as a
diuretic, and removed all his fymptoms.

About three months afterwards he was out upon
a journey, and, after taking cold, was fuddenly
feized with difficulty of breathing, and violent palpitation of his heart: he fent for me, and I ordered
the infufion as before, which very foon removed
his complaints. He is now active and well; but
whenever he takes cold, finds fome return of difficult
breathing, which he foon removes by a dofe or two
of the infufion.

C A S E   XXI.

January 5th. Mrs. M——, Æt. 69. Hydrothorax, (called afthma) afcites and anafarca. I directed an infufion of Fol. Digital. ficcat. three drams
to a pint; a fmall wine glafs to be taken every third
or fourth hour. It made her violently fick, acted
powerfully as a diuretic, fet her breath perfectly at
liberty, and carried off the fwelling of her legs;
when fhe was nearly emptied, fhe became fo languid, that I thought it neceffary to order cordials,
and a large blifter to her back. Mr. Ward, who
attended as her apothecary, tells me fhe had fome
return of her afthma in *June* and *October* following,
which was each time removed by the fame medicine.

CASE

C A S E XXII.

January 11th. Mr. H———, Æt. 59. Afcites
and general anafarca. A large corpulent man, and
a hard drinker: he had repeatedly fuffered under
complaints of this kind, but had been always relieved by the judicious affiftance of Dr. Afh. In
the prefent inftance, however, not finding relief as
ufual from the prefcriptions of my worthy friend,
he fent for me; after examining into his fituation, and
informing myfelf what had been done to relieve him,
I was fatisfied that the Digitalis was the only medicine from which I had any thing to hope. It was
therefore directed; but another patient requiring
my affiftance at a diftance from town, I defired he
would not begin the medicine before I returned,
which would be early on the third day; for I was
well aware of the difficulties before me, and that
he would inevitably fink under too rapid an evacuation of the water. On my return I was informed,
that the preceding evening, as he fat on his chair,
his head funk upon his breaft, and he died.

This cafe, as well as cafe VI. is mentioned with
a view to demonftrate to younger practitioners, how
fudden and unexpected the deaths of dropfical patients fometimes happen, and how cautious we fhould
be in affigning caufes for effects.

C A S E XXIII.

Auguft 31ft. Mr. C———, Æt. 57. Difeafed
vifcera, jaundice, afcites and anafarca. After trying

ing calomel, faline draughts, jallap purges, chryftals
of tartar, pills of gum ammoniac, fquills, and
foap, fal fuccini, eleterium, &c. infufion of Digitalis was directed, which removed all his urgent
fymptoms, and he recovered a pretty good ftate of
health.

C A S E   XXIV.

September 11th. I was defired to vifit Mr. L——,
Æt. 63; a middle fized man; rather thin; not ha-
bitually intemperate; found him in bed, where he
had been for three days. He was in a ftate of furi-
ous infanity, and had been gradually lofing his reafon for ten days before, but was not outrageous the
firft week: his apothecary had given him ten grains
of emetic tartar, a dram of ipecacoanha, and an
ounce of tincture of jallap, in the fpace of a few
hours, which fcarcely made him fick, and only occafioned a ftool or two; upon enquiring into the
ufual ftate of his health, I was told that he had been
troubled with fome difficulty of breathing for thirty
years paft, but for the nine laft years this complaint
had increafed, fo that he was often obliged to fit up
the greater part of the night; and, for the laft year,
the fenfe of fuffocation was fo great, when he lay
down, that he often fat up for a week together. His
father died of an afthma before he was fifty. A few
years ago, at an election, where he drank more
than ufual, his head was affected as now, but in a
flighter degree, and his afthmatic fymptoms vanifhed; and now, notwithftanding he has been feveral
days

days in bed, he feels not the leaſt difficulty in breathing.

Apprehending that the inſanity might be owing to the ſame cauſe which had heretofore occaſioned the aſthma, and that this cauſe was water; I ordered a decoction of the Fol. ſiccat Digital. three drams to half a pint; three ſpoonfuls to be taken every third hour: the fourth doſe made him ſick; the medicine was then ſtopped; the ſickneſs continued at intervals, more or leſs, for four days, during which time he made a great quantity of water, and gradually became more rational. On the fifth day his appetite began to return, and the ſickneſs ceaſed, but the flow of urine ſtill continued.

A week afterwards I ſaw him again, and examined him particularly; his head was then perfectly rational, apetite very good, breath quite eaſy, permitting him to lie down in bed without inconvenience, makes plenty of water, coughs a little, and expectorates freely. He took no other medicine, except a little rhubarb when coſtive.

C A S E   XXV.

September 15th. Mr. J. R——, Æt. 50. Subject to an aſthmatical complaint for more than twenty years, but was this year much worſe than uſual, and ſymptoms of dropſy appeared. In *July* he took G. ammon. ſquill and ſeneka, with infuſ. amarum and foſſil alkaly. In *Auguſt*, infuſum amar.

with

with vin. chalyb. and at bed-time pil. ſtyr. and
ſquill. His complaints increaſing, the ſquill was
puſhed as far as could be borne, but without any
good effect. *September* 15th, an infuſion of Digitalis
was directed, but he died the next morning.

C A S E   XXVI.

September 18th. Mrs. R——, Æt. 30. After a
ſevere child-bearing, found both her legs and thighs
ſwelled to the utmoſt ſtretch of the ſkin. They
looked pale, and almoſt tranſparent. The caſe being ſimilar to that related at No. VIII. I determined
upon a ſimilar method of treatment; but as this patient had an inflammatory ſore throat alſo, I wiſhed
to get that removed firſt, and in three or four days
it was done. I then directed an infuſion of Digitalis, which ſoon increaſed the urinary ſecretion,
and reduced the ſwellings, without any diſturbance
of her ſtomach.

A few days after quitting her bed and coming
down ſtairs, ſome degree of ſwelling in her legs returned, which was removed by calomel, an opening
electuary, and the application of rollers.

C A S E   XXVII.

October 7th. Mr. F——, a little man, with a
ſpine and thorax greatly deformed; for more than
a year paſt had complained of difficult reſpiration,
and a ſenſe of fulneſs about his ſtomach; theſe complaints increaſing, his abdomen gradually enlarged,
and

and a fluctuation in it became perceptible. He had no anafarca, no appearance of difeafed vifcera, and no great paucity of urine. Purges and diuretics of different kinds affording him no relief, my affiftance was defired. After trying fquill medicines without effect, he was ordered to take Pulv. fol. Digital. in fmall dofes. Thefe producing no fenfible effect, the dofes were gradually increafed until naufea was excited; but there was no alteration in the quantity of urine, and confequently no relief to his complaints. I then advifed tapping, but he would not hear of it; however, the diftrefs occafioned by the increafing fulnefs of his belly at length compelled him to fubmit to the operation on the 20th of *November*. It was neceffary to draw off the water again upon the following days:

> *December* the 8th.
> — — 27th.
> 1780. *February* the 4th.
> — — 23d.
> *March* the 9th.

During the intervals, no method I could think of was omitted to prevent the return of the difeafe, but nothing feemed to avail. In the operation of *February* 23d, his ftrength was fo much reduced, that the water was not entirely removed; and on the 9th of March, before his belly was half emptied, notwithftanding the moft judicious application of bandage, his debility was fo great, that it was judged prudent to ftop. After being placed in bed, the faintnefs and ficknefs continued; fevere rigors enfued,

enfued, and violent vomiting; thefe vomitings continued through the night, and in the intervals he
lay in a ftate nearly approaching to fyncope. The
next day I found him with nearly the fame fymptoms, but remarked that the quantity of fluid he
had thrown up was very much more than what he
had taken, and that his abdomen was confiderably
fallen; in the courfe of two or three days more, he
difcharged the whole of the effufed fluid; his ftrength
and appetite gradually returned, and he was in all
refpects much better than he had been before the
laft operation.

Some time afterwards, his belly began to fill
again, and he again applied to me; upon an accurate examination, I judged the quantity of fluid
might then be about four or five quarts. Nature
had pointed out the true method of cure in this
cafe; I therefore ordered him to bed, and directed
ipecacoanha vomits to be given night and morning:
in two or three days the whole of the water was
removed by vomiting, for he never purged, nor
was the quantity of his urine increafed; his appetite and ftrength gradually returned; he never had
any further relapfe, and is now an active healthy
man. I muft leave the reader to make his own re-
flections on this fingular cafe.

CASE

1780.

C A S E   XXVIII.

January 11th. Captain V——, Æt. 42. Had
fuffered much from refiding in hot climates, and
drinking very freely, particularly rum in large quantity. He had tried many phyficians before I faw
him, but nothing relieved him. I found him
greatly emaciated, his countenance of a brownifh
yellow; no appetite, extremely low, diftreffing
fulnefs acrofs his ftomach; legs and thighs greatly
fwollen; pulfe quick, and very feeble; urine in
fmall quantity. As he had evidently only a few
days to live, I ordered him nothing but a folution
of fal diureticus in cinnamon water, flightly acidu-
lated with fyrup of lemons. This medicine effect-
ing no change, and his fymptoms becoming daily
more diftreffing, I directed an infufion of Digitalis.
A few dofes occafioned a copious flow of urine,
without ficknefs or any other difturbance. The medicine was difcontinued; and the next day the urine
continuing to be fecreted very plentifully, he loft
his moft diftreffing complaints, was in great fpirits,
and ate a pretty good dinner. In the evening, as
he was converfing chearfully with fome friends, he
ftooped forwards, fell from his chair, and died in-
ftantly. Had he been in bed, I think there is reafon to believe this fatal fyncope, if fuch it was,
would not have happened.

C   CASE

C A S E   XXIX.

February 6th. Mr. H——, Æt. 63. A corpulent man; had fuffered much from gout, which for
the laft year or two had formed very imperfectly.
He had now fymptoms of water in his cheft, his
belly and his legs. An infufion of Digitalis removed
thefe complaints, and after being confined for the
greater part of the winter, he was well enough to
get abroad again. In the courfe of a month the
dropfical fymptoms returned, and were again removed by the fame medicine. Bitters and tonics
were now occafionally prefcribed, but his debility
gradually increafed, and he died fome time afterwards; but the dropfy never returned.

C A S E   XXX.

February 17th. Mr. D——, Æt. 50. Afcites
and anafarca, with fymptoms of phthifis. He had
been a very hard drinker. The infufum Digitalis
removed his dropfical fymptoms, and he was fufficiently recovered to take a journey; but as the
fpring advanced, the confumptive fymptoms increafed, and he died foon afterwards, perfectly emaciated.

C A S E   XXXI.

March 5th. I was defired to vifit Mrs. H——,
a very delicate woman, who after a fevere lying-in,
had her legs and thighs fwollen to a very great degree;

gree; pale and femi-tranfparent. I found her extremely faint, her pulfe very fmall and flow; vomiting violently, and frequently purging. She was attended by a gentleman who had feen me give the
Digitalis in a fimilar cafe of fwelled legs after a lying-
in (fee Cafe XXVI.) about fix months before. He
had not confidered that this patient was delicate,
the other robuft; nor had he attended to ftop the
exhibition of the medicine when its effects began to
take place. The great diftrefs of her fituation was
evidently owing to the imprudent and unlimited
ufe of the Digitalis. I was very apprehenfive for
her fafety; ordered her cordials and volatiles; a free
fupply of wine, chamomile tea with brandy for
common drink, and blifters. The next day the fituation of things was much the fame, but with all this
difturbance no increafed fecretion of urine. The fame
methods were continued; an opiate ordered at night,
and liniment. volatile upon flannel applied to the
groins, as fhe now complained of great pain in thofe
parts. The third day the naufea was lefs urgent,
the vomitings lefs frequent, the pulfe not fo flow.
Camphorated fpirit, with cauftic volatile alkaly, was
applied to the ftomach, emulfion given for common
drink, and the fame medicines repeated. From
this time, the intervals became gradually longer between the fits of vomiting, the flow of urine increafed, the fwellings fubfided, the appetite returned, and fhe recovered perfectly.

CASE

C A S E   XXXII.

March 16th. Mr. D——, Æt. 70. A paralytic
ſtroke had for ſome weeks paſt impaired the uſe of
his left ſide, and he complained much of his breath,
and of a ſtraitneſs acroſs his ſtomach; at length, an
anaſarca and aſcites appearing, I had no doubt as to
the cauſe of the former ſymptoms; but, upon account of his advanced age, and the paralytic affection, I heſitated to give the Digitalis, and therefore tried the other uſual modes of practice, until
at length his breath would not permit him to lie
down in bed, and his other ſymptoms increaſed ſo
rapidly as to threaten a ſpeedy diſſolution. In this
dilemma I ventured to preſcribe an infuſion of the
Fol. ſiccat. Digital. which preſently excited a copious
flow of urine, and made him very ſick; a ſtrong
infuſion of chamomile flowers, with brandy, relieved
the ſickneſs, but the diuretic effects of the Digitalis
continuing, his dropſy was removed, and his breathing became eaſy. The palſy remained nearly in
the ſame ſtate. He lived until *Auguſt* 1782, and
without any return of the dropſy.

C A S E   XXXIII.

March 18th. Miſs S——, Æt. 5. Hydrocephalus internus. As the caſe did not yield to calomel,
when matters were nearly advanced to extremities,
it occurred to me to try the Infuſum Digitalis; a
few doſes of which were given, but had no ſenſible
effect.

C A S E

C A S E XXXIV.

March 19th. A young lady, foon after the birth of an illegitimate child, became infane. After being near a month under my care, fwellings of her legs, which at firft had been attributed to weaknefs, extended to her thighs and belly; her urine became foul, and fmall in quantity, and the infanity remained nearly the fame. As it had been very difficult to procure evacuations by any means, I ordered half an ounce of Fol. Digital. ficcat. in a pint infufion, and directed two fpoonfuls to be given every two hours: this had the defired effect; the dropfy and the infanity difappeared together, and fhe had afterwards no other medicine but fome aperient pills to take occafionally.

C A S E XXXV.

April 12th. Mr. R——, Æt. 32. For the laft three or four years had had more or lefs of what was confidered as afthma;—it appeared to me Hydrothorax. I directed an infufion of Digitalis, which prefently removed his complaints. In *June* following he had a relapfe, and took two grains of the Pulv. fol. Digit. three times a day, which cured him after taking forty grains, and he has never had a return.

C A S E XXXVI.

May 15th. Mrs. H——, Æt. 40. A fpafmodic afthma, attended with fymptoms of effufion.
An infufion of Digitalis relieved her very confiderably, and fhe lived four years afterwards without
any relapfe.

C A S E XXXVII.

May 26th. R—— B——, Æt. 12. Scrophu-
lous, confumptive, and at length anafarcous. Took
Infuf Digital. without advantage. Died the *July*
following.

C A S E XXXVIII.

June 4th. Mrs. S——, of W——, Æt 49.
Afcites and anafarca. Had taken many medicines;
firft from her apothecary, afterwards by the direction of a very judicious and very celebrated phyfician, but nothing retarded the increafe of the
dropfy. I firft faw her along with the phyfician
mentioned above, on the 14th of *May;* we directed an electuary of chryftals of tartar, and Seltzer
water for common drink; this plan failing, as others
had done before, we ordered the Infuf. Digital. which
in a few days nearly removed the dropfy. I then
left her to the care of her phyfician; but her conftitution was too much impaired to admit of reftoration to health, and I underftand fhe died a few
weeks afterwards.

CASE

C A S E   XXXIX.

June 13th. Mr. P———, Æt. 35. A very hard drinker, was attacked with a fevere hæmoptoe, which was followed by afcites and anafarca. He had every appearance of difeafed vifcera, and his urine was fmall in quantity. The powder and the infufion of Digitalis were given at different times, but without the defired effect. Other medicines were tried, but in vain. Tapping prolonged his exiftence a few weeks, and he died early in the following autumn.

C A S E   XL.

June 27th. Mr. W——, Æt. 37. An apparently afthmatic affection, gradually increafing for three or four years, which not yielding to the ufual remedies, he took the infufion of Digitalis. Two or three dofes made him very fick; but he thought his breathing relieved. After one week he took it again, and was fo much better as to want no other medicine.

In the courfe of the following winter he became hectic, and died confumptive about a year after-wards.

C A S E   XLI.

July 6th. Mr. E——, Æt. 57. Hydrothorax and anafarca; his breath fo fhort that he could not lie.

lie down.   After a trial of fquill, fixed alkaly, and
dulcified fpirit of nitre, I directed Pulv. Digital.
gr. 2, thrice a day.   In four days he was able to
come down ftairs; in three days more no appearance
of difeafe remained ; and under the ufe of aromatics
and fmall dofes of opium, he foon recovered his
ftrength.

C A S E   XLII.

July 7th.   Mifs H——— of T———, Æt. 39.   In
the laft ftage of a phthifis pulmonalis became dropfical.   She took the Digitalis without being relieved.

C A S E   XLIII.

July 9th.   Mrs. F————, Æt. 70.   A chearful, ftrong, healthy woman ; but for a few years
back had experienced a degree of difficult breathing
when in exercife.   In the courfe of the laft year her
legs fwelled, and fhe felt great fulnefs about her
ftomach.   Thefe fymptoms continued increafing
very faft, notwithftanding feveral attempts made by
a very judicious apothecary to relieve her.   The
more regular practitioner failing, fhe had recourfe
to a quack, who I believe plied her very powerfully
with Daphne laureola, or fome draftic purge of that
kind.   I found her greatly reduced in ftrength, her
belly and lower extremities fwollen to an amazing
fize, her urine fmall in quantity, and her appetite
greatly impaired.   For the firft fortnight of my attendance blifters were applied, folution of fixed
alkaly, decoction of feneka with vitriolic æther,

chryftals

chryftals of tartar, fquill and cordial medicines were fucceffively exhibited, but with no advantage. I then directed Pulv. Fol. Digital, two grains every four hours. After taking eighteen grains, the urine began to increafe. The medicine was then ftopped. The difcharge of urine continued to increafe, and in five or fix days the whole of the dropfical water paffed off, without any difturbance to the ftomach or bowels. As the diftenfion of the belly had been very great, a fwathe was applied, and drawn gradually tighter as the water was evacuated. As no pains were fpared to prevent the return of the dropfy, and as the beft means I could devife proved unequal to my wifhes, both in this and in fome other cafes, I fhall take the liberty to point out the methods I tried at different times in as concife a manner as poffible, for the knowledge of what will not do, may fometimes affift us to difcover what will.

1780.

July 18th. Infufum amarum, fteel, Seltzer water.

September 22d. Neutral faline draughts, with tinct. canthar.

26th. Pills of foap, garlic and millepedes.

30th. The fame pills, with infufum amarum.

October 11th. Pills of aloes, affafetida, and fal martis, in the day-time, and mercury rubbed down, at night.

December 21ft. The accumulation of water now required a repetition of the Digitalis. It was directed in infufion, a dram and half to eight ounces, and an ounce and half given every fourth hour,

until

until its effects began to appear.   The water was
foon carried off.

30th.   Sal diuretic. twice a day.   To eat preferved
garlic frequently.

1781.

February 1ft.   Pills of calomel, fquill and gum ammoniac.

3d.   Infufion of Digitalis repeated, and after the
water was carried off, Dover's powder was tried
as a fudorific.

March 18th.   Infuf. Digital. repeated.

26th.   Pills of fal martis and aromatic fpecies, with
infufum amarum.

May 5th.   Being feverifh; James's powder and
faline draughts.

10th.   Laudanum every night, and an opening
tincture to obviate coftivenefs.

24th.   Infuf. Digitalis, one ounce only every fourth
hour, which foon procured a perfect evacuation
of the water.

Auguft 11th.   Infuf. Digitalis.

October 19th.   An emetic, and fol. Cicut. pulv.
ten grains every fix hours.

November 8th.   A mercurial bolus at bed-time.

16th.   Infuf. Digitalis.

December 23d. An emetic—Pills of feneka and gum
ammoniac—Vitriolic acid in every thing fhe
drinks.

25th.   Squill united to fmall dofes of opium.

1782.

January 2d.   A troublefome cough—Syrup of garlic and oxymel of fquills.   A blifter to the back.

4th.   Tincture

4th. Tincture of cantharides and paregoric elixir.

28th. Infuf. Digitalis, half an ounce every morning, and one ounce every night, was now fufficient to empty her.

March 26th. Infuf. Digitalis; and when emptied, vitriol of copper twice a day.

April 1ft. A cordial mixture for occafional ufe.

Two months afterwards a purging came on, which every now and then returned, inducing great weaknefs—her appetite failed, and fhe died in in *July*.

INTERVALS.

From *July* 9th, 1780, to *December* 21ft, 171 days.

From *December* 21ft to *February* 3d, 1781, 34 days.

From *February* 3d to *March* 18th, 44 days.

From *March* 18th to *May* 24th, 66 days.

From *May* 24th to *Auguft* 11th, 79 days.

From *Auguft* 11th to *November* 16th, 98 days.

From *November* 16th to *January* 28th, 1782, 74 days.

From *January* 28th to *March* 26th, 57 days.

None of the accumulations of water were at all equal to that which exifted when I firft faw her, for finding fo eafy a mode of relief, fhe became impatient under a fmall degree of preffure, and often infifted upon taking her medicine fooner than I thought it neceffary. After the 26th of *March* the degree of effufion was inconfiderable, and at the time of her death very trifling, being probably carried off by the diarrhoea.

CASE

C A S E   XLIV

July 12th. Mr. H——, of A——, Æt. 60. In the laft ftage of a life hurried to a termination by free living, dropfical fymptoms became the moft diftreffing. He wifhed to take the Digitalis. It was given, but afforded no relief.

C A S E   XLV.

July 13th. Mr. S——, Æt. 49. Afthma, or rather hydrothorax, anafarca, and fymptoms of a difeafed liver. He was directed to take two grains of Pulv. fol. Digital. every two hours, until it produced fome effect. It foon removed the dropfical and afthmatic affections, and fteel, with Seltzer water, reftored him to health.

C A S E   XLVI.

Auguft 6th. Mr. L——, Æt. 35. Afcites and anafarca. Pulv. Digital. grains three, repeated every fourth hour, until he had taken two fcruples, removed every appearance of dropfy in a few days. He was then directed to take folution of merc. fublimat. and foon recovered his health and ftrength.

C A S E   XLVII.

Auguft 16th. Mr. G——, of W——, Æt. 86. Afthma of many years duration, and lately an incipient anafarca, with a paucity of urine. He had never lived intemperately, was of a chearful difpofition, and very fenfible: for fome years back had

loft

loft all relifh for animal food, and his only fupport
had been an ounce or two of bread and cheefe, or
a fmall flice of feed-cake, with three or four pints
of mild aie, in the twenty-four hours. After trying chryftals of tartar, fixed alkaly, fquills, &c. I
directed three grains of Pulv. fol. Digital. made
into pills, with G. ammoniac, to be given every fix
hours; this prefently occafioned copious difcharges
of urine, removed his fwellings, and reftored him
to his ufual ftandard of health.

C A S E   XLVIII.

Auguft 17th. T—— B——, Efq. of K——,
Æt. 46. Jaundice, dropfy, and great hardnefs in
the region of the liver. Infufion of Digitalis carried off all the effufion, and afterwards a courfe of
deobftruent and tonic medicines removed his other
complaints.

C A S E   XLIX.

Auguft 23d. Mr. C——, Æt. 58. (The perfon
mentioned at Cafe XXIII.) He had continued free
from dropfy until within the laft fix weeks; his appetite was now totally gone, his ftrength extremely
reduced, and the yellow of his jaundice changed to a
blackifh hue. The Digitalis was now tried in vain,
and he died fhortly afterwards.

C A S E   L.

Auguft 24th. Mrs. W——, Æt. 39. Anafarcous legs and fymptoms of hydrothorax, confequent

to

to a tertian ague Three grains of Pulv. Digitalis, given every fourth hour, occafioned a very copious flow of urine, and fhe got well without any other medicine.

C A S E LI.

Auguft 28th. Mr. J—— H——, Æt. 27. In confequence of very free living, had an afcites and fwelled legs. I ordered him to take two grains of Fol. Digital. pulv. every two hours, until it produced fome effect; a few dofes caufed a plentiful fecretion of urine, but no ficknefs, or purging: in fix days the fwellings difappeared, and he has fince remained in good health.

C A S E LII.

September 27th. Mr. S——, Æt. 45. Had been long in an ill ftate of health, from what had been fuppofed an irregular gout, was greatly emaciated, had a fallow complexion, no appetite, coftive bowels, quick and feeble pulfe. The caufe of his complaints was involved in obfcurity; but I fufpected the poifon of lead, and was ftrengthened in this fufpicion, upon finding his wife had likewife ill health, and, at times, fevere attacks of colic; but the anfwers to my enquiries feemed to prove my fufpicions fruitlefs, and, amongft other things, I was told the pump was of wood. He had lately fuffered extremely from difficult breathing, which I thought owing to anafarcous lungs; there was alfo a flight degree of pale fwelling in his legs. Pulv.
fol.

fol. Digital. made into pills, with gum ammoniac and aromatic fpecies, foon relieved his breathing. Attempts were then made to affift him in other re-fpects, but with little good effect, and fome months afterwards he died, with every appearance of a worn out conftitution.

About two years after this gentleman's death, I was talking to a pump-maker, who, in the courfe of con-verfation, mentioned the corrofion of leaden pumps, by fome of the water in this town, and inftanced that at the houfe of Mr. S——, which he had re-placed with a wooden one about three years before. The lead, he faid, was eaten away, fo as to be very thin in fome places, and full of holes in others;—this accidental information explained the myftery.

The deleterious effects of lead feem to be confi-derably modified by the conftitution of the patient; for in fome families only one or two individuals fhall fuffer from it, whilft the reft receive it with impunity. In the fpring of the year 1776, I was defired to vifit Mrs. H——, of S—— Park, who had repeatedly been attacked with painful colics, and had fuffered much from infuperable coftivenefs; I fufpected lead to be the caufe of her complaints, but was unable to trace by what means it was taken. She was relieved by the ufual methods; but, a few months afterwards, I was defired to fee her again: her fufferings were the fame as before, and notwith-ftanding every precaution to guard againft coftive-nefs, fhe was never in perfect health, and feldom

escaped

efcaped fevere attacks twice or thrice in a year; fhe had alfo frequent pains in her joints. I could not find any traces of fimilar complaints either in Mr. H——, the children, or the fervants Mrs. H—— was a water drinker, and feldom tafted any fermented liquor. The pump was of wood, as I had been informed upon my firft vifit. Her health continued nearly in the fame ftate for two or three years more, but fhe always found herfelf better if fhe left her own houfe for any length of time. At length it occurred to me, that though the pump was a wooden one, the pifton might work in lead. I therefore ordered the pump rods to be drawn up, and upon examination with a magnifying glafs, found the leather of the pifton covered with an infinite number of very minute fhining particles of lead. Perhaps in this inftance the metal was fo minutely divided by abrafion, as to be mechanically fufpended in the water. The lady was directed to drink the water of a fpring, and never to fwallow that from the pump. The event confirmed my fufpicions, for fhe gradually recovered a good ftate of health, loft the obftinate coftivenefs, and has never to this day had any attack of the colic.

C A S E LIII.

September 28th. Mrs. J——, Æt. 70. Afcites and very thick anafarcous legs and thighs, total lofs of ftrength and appetite. Infufion of Digitalis was given, but, as had been prognofticated, with no good effect.

C A S E

C A S E   LIV

September 30th. Mr. A——, Æt. 57. A ſtrong
man; hydrothorax and ſwelled legs ; in other reſ-
pects not unhealthful. He was directed to take two
grains of the Pulv. fol. Digit. made into a pill with
gum ammoniac. Forty grains thus taken at intervals,
effected a cure by increaſing the quantity of urine,
and he has had no relapſe.

C A S E   LV

November 2d. Mr. P—— of T——, Æt. 42. A
very ſtrong man, drank a great quantity of ſtrong
ale, and was much expoſed to alterations of heat
and cold.   About the end of ſummer found himſelf
ſhort winded, and loſt his appetite. The dyſpnœa
gradually increaſed, he got a moſt diſtreſſing ſenſe
of tightneſs acroſs his ſtomach, his urine was little,
and high coloured, and his legs began to ſwell; his
pulſe ſlender and feeble. From the 20th of *September* I frequently ſaw him, and obſerved a gradual
and regular increaſe of all his complaints, notwithſtanding the uſe of the moſt powerful medicines I
could preſcribe. He took chryſtals of tartar, ſeneka,
gum ammoniac, ſaline draughts, emetics, tinct. of
cantharides, ſpirits of nitre dulcified, ſquills in all
forms, volatile alkaly, calomel, Dover's powder,
&c. Bliſters and draſtic purgatives were tried, in-
terpoſing ſalt of ſteel and gentian. I had all along
felt a reluctance to preſcribe the Digitalis in this
caſe, from a perſuaſion that it would not ſucceed.

D   At

At length I was compelled to it, and directed one grain to be given every two hours until it should excite nausea. This it did; but, as I expected, it did no more. The reason of this belief will be mentioned hereafter. Five days after this last trial I gave him affafetida in large quantity, flattered by a hope that his extreme sufferings from the state of his respiration, might perhaps arise in part from spasm, but my hopes were in vain. I now thought of using an infusion of tobacco, and prescribed the following:

R. Fol. Nicotian. incif. ʒii.
    Aq. bull. ℔fs.
    Sp. Vini rectif. ʒi digere per horam.

I directed a spoonful of this to be given every two hours until it should vomit. This medicine had no better effect than the former ones, and he died some days afterwards.

C A S E   LVI.

November 6th. Mr. H——, Æt. 47. In the last stage of a phthifis pulmonalis, suffered much from dyspnœa, and anafarca. Squill medicines gave no relief. Digitalis in pills, with gum ammon. purged him, but opium being added, that effect ceased, and he continued to be relieved by them as long as he lived.

C A S E

CASE LVII.

November 16th. Mrs. F————, Æt. 53. In *Auguſt* laſt was ſuddenly ſeized with epileptic fits, which continued to recur at uncertain intervals. Her belly had long been larger than natural, but without any perceptible fluctuation. Her legs and thighs ſwelled very conſiderably the beginning of this month, and now there was evidently water in the abdomen. The medicines hitherto in vain directed againſt the epileptic attacks, were now ſuſpended, and two grains of the Pulv. fol. Digital. directed to be taken every ſix hours. The effects were moſt favourable, and the dropſical ſymptoms were ſoon removed by copious urinary diſcharges.

The attacks of epilepſy ceaſed ſoon afterwards. In *February*, 1781, there was ſome return of the ſwellings, which were ſoon removed, and ſhe now enjoys very good health. Does not the narrative of this caſe throw light upon the nature of the epilepſy which ſometimes attacks women, ſoon after the ceſſation of the menſtrual flux?

1781.

CASE LVIII.

January 1ſt. Mrs. G————, of H————, Æt. 62. Aſcites and very large hard legs. After trying various medicines, under the direction of a very able phyſician, I ordered her to take one grain of Pulv.

Digital.

Digital. every fix hours, but it produced no effect. Other Medicines were then tried to as little pur-pofe. About the end of *February*, I directed an infufion of the Fol. Digital. but with no better fuc-cefs. Other methods were thought of, but none proved efficacious, and fhe died a few weeks after-wards.

C A S E LIX.

January 3d. Mrs. B———, Æt. 53. Afcites, anafarca, and jaundice. After a purge of calomel and jallap, was ordered the Infufion of Digitalis: it acted kindly as a diuretic, and greatly reduced her fwellings. Other medicines were then adminifter-ed, with a view to her other complaints, but to no purpofe, and fhe died about a month afterwards.

C A S E LX.

January 14th. Mr. B———, of D———. Jaun-dice and afcites, the confequences of great intem-perance. Extremely emaciated; his tongue and fauces covered with apthous crufts, and his appetite gone. He firft took tincture of cantharides with infufum amarum, then vitriolic falts, and various other medicines without relief; Infufum Digitalis was given afterwards, but was equally unfuccefsful.

C A S E LXI.

February 2d. I was defired by the late learned and ingenious Dr. Groome, to vifit Mifs S———, a

young

young lady in the laſt ſtate of emaciation from a dropſy. Every probable means to relieve her had been attempted by Dr. Groome, but to no pur-poſe; and ſhe had undergone the operation of the paracenteſis repeatedly. The Doctor knew, he ſaid, that I had cured many caſes of dropſy, by the Di-gitalis, after other more uſual methods had been attempted without ſucceſs, and he wiſhed this lady to try that medicine under my direction; after exa-mining the patient, and enquiring into the hiſtory of the diſeaſe, I was ſatisfied that the dropſy was encyſted, and that no medicine could avail. The Digitalis, however, was directed, and ſhe took it, but without advantage. She had determined not to be tapped again, and neither perſuaſion, nor diſtreſs from the diſtenſion, could prevail upon her: I at length propoſed to make an opening into the ſac, by means of a cauſtic, which was done under the judicious ma-nagement of Mr. Wainwright, ſurgeon, at Dudley. The water was evacuated without any accident, and the patient afterwards let it out herſelf from time to time as the preſſure of it became troubleſome, un-til ſhe died at length perfectly exhauſted.

Query. Is there not a probability that this me-thod, aſſiſted by bandage, might be uſed ſo as to effect a cure, in the earlier ſtages of ovarium dropſy?

C A S E   LXII.

February 27th. Mrs. O——, of T——, Æt. 52, with a conſtitution worn out by various complicated

diſorders

diforders, at length became dropfical. The Digitalis was given in fmall dofes, in hopes of temporary
benefit, and it did not fail to fulfil our expectations.

C A S E LXIII.

March 16th. Mrs. P——, Æt. 47. Great debility, pale countenance, lofs of appetite, legs fwelled,
urine in fmall quantity. A dram of Fol. ficcat. Digital. in a half pint infufion was ordered, and an
ounce of this infufion directed to be taken every
morning. Myrrh and fteel were given at intervals.
Her urine foon increafed, and the fymptoms of
dropfy difappeared.

C A S E LXIV.

March 18th. Mr. W——, in the laft ftage
of a pulmonary confumption became dropfical. The
Digitalis was given, but without any good effect.

C A S E LXV.

April 6th. Mr. B——, Æt. 63. For fome
years back had complained of being afthmatical,
and was not without fufpicion of difeafed vifcera. The laft winter he had been moftly confined
to his houfe; became dropfical, loft his appetite,
and his fkin and eyes turned yellow. By the ufe
of medicines of the deobftruent clafs he became lefs
difcoloured, and the hardnefs about his ftomach
feemed to yield; but the afcites and anafarcous
fymptoms increafed fo as to opprefs his breathing
exceed-

exceedingly. Alkaline falts, and other diuretics failing of their effects, I ordered him to take an infuf. of Digitalis. It operated fo powerfully that it be﹣ came neceſſary to fupport him with cordials and blifters, but it freed him from the dropfy, and his breath became quite eafy. He then took foap, rhubarb, tartar of vitriol, and fteel, and gradually attained a good ftate of health, which he ftill continues to enjoy.

C A S E   LXVI.

April 8th. Mr. B——, Æt. 60. A corpulent man, with a ftone in his bladder, from which at times his fufferings are extreme. He had been affected with what was fuppofed to be an afthma, for feveral years by fits, but through the laft winter his breath had been much worfe than ufual; univerfal anafarca came on, and foon afterwards an afcites. Now his urine was fmall in quantity and much faturated, the dyfuria was more dreadful than ever; his breath would not allow him to lie in bed, nor would the dyfuria permit him to fleep; in this diftrefsful fituation, after having ufed other medicines to little purpofe, I directed an infufion of Digitalis to be given. When the quantity of urine became more plentiful, the pain from his ftone grew eafier; in a few days the dropfy and afthma difappeared, and he foon regained his ufual ftrength and health. Every year fince, there has been a tendency to a return of thefe complaints, but he has recourfe to the infufion, and immediately removes them.

CASE

C A S E   LXVII.

April 24th. Mr. M——, of C—— Æt. 57.
Afthma, anafarca, jaundice, and great hardnefs and
ftraitnefs acrofs the region of the ftomach. After a
free exhibition of neutral draughts, alkaline falt,
&c. the dropfy and difficult breathing remaining the
fame, he took Infufum Digitalis, which removed
thofe complaints. He never loft the hardnefs about
his ftomach, but enjoyed very tolerable health for
three years afterwards, without any return of the
dropfy.

C A S E   LXVIII.

April 25th. Mrs. J——, Æt. 42. Phthifis pulmonalis and anafarcous legs and thighs. She took
the Infufum Digitalis without effect. Myrrh and
fteel, with fixed alkaly, were then ordered, but to
no purpofe.

C A S E   LXIX.

May 1ft. Mafter W——, of St——, Æt. 6.
I found him with every fymptom of hydrocephalus
internus. As it was yet early in the difeafe, in confequence of ideas which will be mentioned hereafter, I directed fix ounces of blood to be immediately taken from the arm; the temporal artery to
be opened the fucceeding day; the head to be fha-
ven, and fix pints of cold water to be poured upon
it every fourth hour, and two fcruples of ftrong mercurial

curial ointment to be rubbed into the legs every day. Five days afterwards, finding the febrile fymptoms very much abated, and judging the remaining difeafe to be the effect of effufion, I directed a fcruple of Fol. Digital. ficcat. to be infufed in three ounces of water, and a table fpoonful of the infufion to be given every third or fourth hour, until its action fhould be fomeway fenfible.   The effect was, an increafed fecretion of urine; and the patient foon recovered.

C A S E   LXX.

May 3d. Mrs. B———, Æt. 59. Afcites and anafarca, with ftrong fymptoms of difeafed vifcera. Infufum Digitalis was at firft prefcribed, and prefently removed the dropfy. She was then put upon faline draughts and calomel. After fome time fhe became feverifh: the fever proved intermittent, and was cured by the bark.

C A S E   LXXI.

May 3d. Mr. S———, Æt. 48. A ftrong man, who had lived intemperately. For fome time paft his breath had been very fhort, his legs fwollen towards evening, and his urine fmall in quantity. Eight ounces of the Infuf. Digitalis caufed a confiderable flow of urine; his complaints gradually vanifhed, and did not return.

C A S E

C A S E   LXXII.

May 24th. Joſeph B——, Æt. 50. Aſcites, anaſarca, and jaundice, from intemperate living. Infuſion of Digitalis produced nauſea, and lowered the
frequency of the pulſe; but had no other ſenſible effects. His diſorder continued to increaſe, and killed
him about two months afterwards.

C A S E   LXXIII.

June 29th. Mr. B——, Æt. 60. A hard drinker;
afflicted with aſthma, jaundice, and dropſy. His
appetite gone; his water foul and in ſmall quantity.
Neutral ſaline mixture, chryſtals of tartar, vinum
chalybeat and other medicines had been preſcribed
to little advantage. Infuſion of Fol. Digitalis acted
powerfully as a diuretic, and removed the moſt urgent of his complaints, viz. the dropſical and aſthmatical ſymptoms.

The following winter his breathing grew bad again,
his appetite totally failed, and he died, but without
any return of the aſcites.

C A S E   LXXIV.

June 29th. Mr. A——, Æt. 58. Kept a public
houſe and drank very hard. He had ſymptoms of
diſeaſed viſcera, jaundice, aſcites, and anaſarca. After taking various deobſtruents and diuretics, to no
purpoſe, he was ordered the Infuſion of Digitalis:
a few

a few dofes occafioned a plentiful flow of urine, relieved his breath, and reduced his fwellings; but,
on account of his great weaknefs, it was judged imprudent to urge the medicine to the entire evacuation of the water. He was fo much relieved as to
be able to come down ftairs and to walk about, but
his want of appetite and jaundice continuing, and
his debility increafing, he died in about two
months.

C A S E   LXXV.

July 18th. Mrs. B——, Æt. 46. A little woman, and very much deformed. Afthmatical for
many years. For feveral months paft had been worfe
than ufual; appetite totally gone, legs fwollen,
fenfe of great fulnefs about her ftomach, countenance fallen, lips livid, could not lie down.

The ufual modes of practice failing, the Digitalis
was tried, but with no better fuccefs, and in about a
month fhe died; not without fufpicion of her death
having been accelerated a few days, by her taking
half a grain of opium. This may be a caution to
young practitioners to be careful how they venture
upon even fmall dofes of opium in fuch conftitutions, however much they may be urged by the patient to prefcribe fomething that may procure a little
reft and eafe.

C A S E   LXXVI.

August 12th. Mr. L——, Æt. 65, the perfon whofe Cafe is recorded at No. XXIV, had a return of his infanity, after near two years perfect health. He was extremely reduced when I faw him, and the medicine which cured him before was now adminiftered without effect, for his weaknefs was fuch that I did not dare to urge it.

C A S E   LXXVII.

September 10th. Mr. V——, of S——, Æt. 47. A man of ftrong fibre, and the remains of a florid complexion. His difeafe an afcites and fwelled legs, the confequence of a very free courfe of life; he had been once tapped, and taken much medicine before I faw him. The Digitalis was now directed: it lowered his pulfe, but did not prove diuretic. He returned home, and foon after was tapped again, but furvived the operation only a few hours.

C A S E   LXXVIII.

September 25th. Mr. O——, of M——, Æt. 63. Very painful and general fwellings in all his limbs, which had confined him moftly to his bed fince the preceding winter; the fwellings were uniform, tenfe, and refifting, but the fkin not difcoloured. After trying guiacum and Dover's powder without advantage, I directed Infufion of Digitalis. It acted on the kidneys, but did not relieve him. It is not

eafy

eafy to fay what the difeafe was, and the patient living at a diftance, I never learnt the future progrefs or termination of it.

CASE LXXIX.

September 26th. Mr. D——, Æt. 42, a very fenfible and judicious furgeon at B——, in Staffordfhire, laboured under afcites and very large anafarcous legs, together with indubitable fymptoms of difeafed vifcera. Having tried the ufual diuretics to no purpofe, I directed a fcruple of Fol. Digital ficcat. in a four ounce infufion, a table fpoonful to be taken twice a day. The fecond bottle wholly removed his dropfy, which never returned.

CASE LXXX.

September 27th. Mrs. E——, Æt. 42. A fat fedentary woman; after a long illnefs, very indiftinctly marked; had fymptoms of enlarged liver and dropfy. In this cafe I was happy in the affiftance of Dr. Afh. Digitalis was once exhibited in fmall dofes, but to no better purpofe than many other medicines. She fuffered great pain in the abdomen for feveral weeks, and after her death, the liver, fpleen, and kidneys were found of a pale colour, and very greatly enlarged, but the quantity of effufed fluid in the cavity was not more than a pint.

CASE

C A S E   LXXXI.

October 28th. Mr. B——, Æt. 33. Had drank an immenfe quantity of mild ale, and was now become dropfical. He was a lufty man, of a pale complexion: his belly large, and his legs and thighs fwollen to an enormous fize. I directed the Infufion of Digitalis, which in ten days completely emptied him. He was then put upon the ufe of fteel and bitters, and directed to live temperately, which I believe he did, for I faw him two years afterwards in perfect health.

C A S E   LXXXII.

November 14th. Mr. W——, of T——, Æt. 49. A lufty man, with an afthma and anafarca. He had taken feveral medicines by the direction of a very judicious apothecary, but not getting relief as he had been accuftomed to do in former years, he came under my direction. For the fpace of a month I tried to relieve him by fixed alkaly, feneka, Dovers powder, gum ammoniac, fquill, &c. but without effect. I then directed Infufion of Digitalis, which foon increafed the flow of urine without exciting naufea, and in a few days removed all his complaints.

CASE

1782.

C A S E LXXXIII.

January 23d. Mr. Q——, Æt. 74. A ſtone in his bladder for many years; dropſical for the laſt three months. Had taken at different times ſoap with ſquill and gum ammoniac; ſoap lees; chryſtals of tartar, oil of juniper, ſeneka, jallap, &c. but the dropſical ſymptoms ſtill increaſed, and the dyſuria from the ſtone became very urgent. I now directed a dram of the Fol. Digit. ſiccat. in a half pint infuſion, half an ounce to be given every ſix hours. This preſently relieved the dyſuria, and ſoon removed the dropſy, without any diſturbance to his ſyſtem.

C A S E LXXXIV.

January 27th. Mr. D——, Æt. 86. The debility of age and dropſical legs had long oppreſſed him. A few weeks before his death his breathing became very ſhort, he could not lie down in bed, and his urine was ſmall in quantity. A wine glaſs of a weak Infuſion of Digitalis, warmed with aromatics, was ordered to be taken twice a day. It afforded a temporary relief, but he did not long ſurvive.

C A S E LXXXV.

January 28th. Mr. D——, Æt. 35. A publican and a hard drinker. Aſcites, anaſarca, diſeaſed
viſcera

vifcera, and flight attacks of hæmoptoe. A dram
of Fol. Digital. ficc. in a half pint infufion, of which
one ounce was given night and morning, proved
diuretic and removed his dropfy.   He then took
medicines calculated to relieve his other complaints.
The dropfy did not return during my attendance
upon him, which was three or four weeks. A quack
then undertook to cure him with blue vitriol vomits,
but as I am informed, he prefently funk under that
rough treatment.

C A S E LXXXVI.

January 29th. Mrs. O——, of D——, Æt. 53.
A conftant and diftreffing palpitation of her heart,
with great debility.   From a degree of anafarca in
her legs I was led to fufpect effufion in the Pericardium, and therefore directed Digitalis, but it produced no benefit. She then took various other medicines with the fame want of fuccefs, and about
ten months afterwards died fuddenly.

C A S E   LXXXVII.

January 31ft. Mr. T——, of A——, Æt. 81.
Great difficulty of breathing, fo that he had not
lain in bed for the laft fix weeks, and fome fwelling in his legs. Thefe complaints were fubfequent to a
very fevere cold, and he had ftill a troublefome
cough. He told me that at his age he did not look for
a cure, but fhould be glad of relief, if it could be
obtained without taking much medicine. I directed
an Infufion of Digitalis, a dram to eight ounces,

one

one fpoonful to be taken every morning, and two at night. He only took this quantity; for in four days he could lie down, and foon afterwards quitted his chamber. In a month he had a return of his complaints, and was relieved as before.

C A S E LXXXVIII.

January 31ft. Mrs. J——, of S——, Æt. 67. A lufty woman, of a florid complexion, large belly, and very thick legs. She had been kept alive for fome years by the difcharge from ulcers in her legs; but the fores now put on a very difagreeable livid appearance, her belly grew ftill larger, her breath fhort, her pulfe feeble, and fhe could not take nou-rifhment. Several medicines having been given in vain, the Digitalis was tried, but with no better ef-fect; and in about a month fhe died.

C A S E LXXXIX.

February 2d. Mr. B——, Æt. 73. An univer-fal dropfy. He took various medicines, and Digi-talis in fmall dofes, but without any good effect.

C A S E XC.

February 24th. Mafter M——, of W——, Æt. 10. An epilepfy of fome years continuance, which had never been interrupted by any of the various methods tried for his relief. The Digitalis was given for a few days but as he lived at a diftance, fo that I could not attend to its effects, he only took one

E half

half pint infusion, which made no alteration in his complaint.

C A S E   XCI.

March 6th. Mr. H——, Æt. 62. A very hard drinker, and had twice had attacks of apoplexy. He had now an afcites, was anafarcous, and had every appearance of a difeafed liver. Small dofes of calomel, Dover's powder, infufum amarum, and fal fodæ palliated his fymptoms for a while; thefe failing; blifters, fquills, and cordials were given without effect. A weak Infufion of Digitalis, well aromatifed, was then directed to be given in fmall dofes. It rather feemed to check than to increafe the fecretion of urine, and foon produced ficknefs. Failing in its ufual effect, the medicine was no longer continued; but every thing that was tried proved equally inefficacious, and he did not long furvive.

C A S E   XCII.

May 10th. Mrs. P——, Æt. 40. Spafmodic afthma of many years continuance, which had frequently been relieved by ammoniacum, fquills, &c. but thefe now failing in their wonted effects, an Infuf. of Fol. Digitalis was tried, but it feemed rather to increafe than relieve her fymptoms.

C A S E   XCIII.

May 22d. Mr. O——, of B——, Æt. 61. A very large man, and a free liver; after an attack of
hemi-

hemiplegia early in the fpring, from which he only partially recovered, became dropfical. The dropfy occupied both legs and thighs, and the arm of the affected fide. I directed an Infufion of Digitalis in fmall dofes, fo as not to affect his ftomach. The fwellings gradually fubfided, and in the courfe of the fummer he recovered perfectly from the palfy.

C A S E XCIV.

July 5th. Mr. C——, of W——, Æt. 28. Had drank very freely both of ale and fpirits; and in confequence had an afcites, very large legs, and great fulnefs about the ftomach. He was ordered to take the Infufion of Digitalis night and morning for a few days, and then to keep his bowels open with chryftals of tartar. The firft half pint of infufion relieved him greatly; after an interval of a fortnight it was repeated, and he got well without any other medicine, only continuing the chryftals of tartar occafionally. I forgot to mention that this gentleman, before I faw him, had been for two months under the care of a very celebrated phyfician, by whofe direction he had taken mercurials, bitters, fquills, alkaline falts, and other things, but without much advantage.

C A S E XCV.

March 6th. Mrs. W——, Æt. 36. In the laft ftage of a pulmonary confumption, took the Infuf. Digitalis, but without any advantage.

C A S E   XCVI.

Auguſt 20th. Mr. P——, Æt. 43. In the year
1781 he had a fevere peripneumony, from which
he recovered with difficulty. At the date of this,
when he firſt conſulted me, the ſymptoms of hydrothorax were pretty obvious. I directed a purge,
and then the Infuſum Digitalis, three drams to
half a pint, one ounce to be taken every four hours.
It made him ſick, and occaſioned a copious diſcharge
of urine. His complaints immediately vaniſhed,
and he remains in perfect health.

C A S E   XCVII.

September 24th. Mrs. R——, of B——, Æt. 35,
the mother of many children. After her laſt lying
in, three months ago, had that kind of ſwelling in
one of her legs which is mentioned at No. VIII.
XXVI, and XXXI. A conſiderable degree of ſwelling ſtill remained; the limb was heavy to her feel-
ing, and not devoid of pain. I directed a bolus of
five grains of Pulv. Digitalis, and twenty-five of
crude quickſilver rubbed down, with conſerve of cy-
noſbat. to be taken at bed-time, and afterwards an
Infuſion of red bark and Fol. Digitalis to be taken
twice a day. There was half an ounce of bark and
half a dram of the leaves in a pint infuſion: the
doſe two ounces.

The leg ſoon began to mend, and two pints of the
infuſion finiſhed the cure.

CASE

C A S E XCVIII.

September 25th. Mr. R———, Æt. 60. Com‑
plained to me of a ficknefs after eating, and for
fome weeks paft he had thrown up all his food, foon
after he had fwallowed it. He had taken various
medicines, but found benefit from none, and had
tried various kinds of diet. He was now very thin
and weak; but had a good appetite. As feveral
very probable methods had been prefcribed, and as
the ufual fymptoms of organic difeafe were abfent,
I determined to give him a fpoonful of the Infufion
of Digitalis twice a day; made by digefting two
drams of the dried leaves in half a pint of cinnamon
water. From the time he began to take this medi‑
cine he fuffered no return of his complaint, and
foon recovered his flefh and his ftrength.

It fhould be obferved, that I had frequently feen
the Digitalis remove ficknefs, though prefcribed for
very different complaints.

C A S E XCIX.

September 30th. Mrs. A———, Æt. 38. Hydro‑
thorax and anafarca. Her cheft was very confider‑
ably deformed. One half pint of the Digitalis In‑
fufion entirely cured her.

C A S E C.

September 30th. Mr. R——, of W——, Æt. 47. Hydrothorax and anasarca. An Infusion of Digitalis was directed, and after the expected effects from that should take place, sixty drops of tincture of cantharides twice a day. As he was costive, pills of aloes and steel were ordered to be taken occafionally.

This plan fucceeded perfectly. About a month afterwards he had fome rheumatic affections, which were removed by guiacum.

C A S E CI.

October 2d. Mrs. R——, Æt. 60. Difeafed viscera; afcites and anasarca. Had taken various deobstruent and diuretic medicines to little purpose. The Digitalis brought on a naufea and languor, but had no effect on the kidneys.

C A S E CII.

October 12th. Mr. R——, Æt. 41. A publican, and a hard drinker. His legs and belly greatly fwollen; appetite gone, countenance yellow, breath very fhort, and cough troublefome. After a vomit I gave him calomel, faline draughts, steel and bitters, &c. He had taken the more ufual diuretics before I faw him. As the dropfical fymptoms increafed, I changed his medicines for pills made of

soap

foap, containing two grains of Pulv. fol. Digital. in
each dofe, and, as he was coftive, two grains of
jallap. He took them twice a day, and in a week
was free from every appearance of dropfy. The
jaundice foon afterwards vanifhed, and tonics reftored him to perfect health.

C A S E   CIII.

October 12th. Mr. B——, Æt. 39. Kept a public houfe, drank very freely, and became dropfical;
he complained alfo of rheumatic pains. I directed
Infufion of Digitalis, half an ounce twice a day.
In eight days the fwellings in his legs and the fulnefs about his ftomach difappeared. His rheumatic
affections were cured by the ufual methods.

C A S E   CIV.

October 22d. Mafter B——, Æt. 3. Afcites and
univerfal anafarca. Half a grain of Fol. Digital.
ficcat. given every fix hours, produced no effect;
probably the medicine was wafted in giving. An
infufion of the dried leaf was then tried, a dram to
four ounces, two tea fpoonfuls for a dofe; this foon
increafed the flow of urine to a very great degree,
and he got perfectly well.

C A S E   CV.

October 30th. Mr. G——, of W——, Æt. 88.
The gentleman mentioned in No. XLVII. His
complaints and manner of living the fame as there
mentioned.

mentioned. I ordered an Infufion of the Digitalis, a dram and half to half a pint; one ounce to be taken twice a day; which cured him in a fhort time.

On *March* the 23d, 1784, he fent for me again. His complaints were the fame, but he was much more feeble.   On this account I directed a dram of the Fol. Digitalis to be infufed for a night in four ounces of fpirituous cinnamon water, a fpoonful to be taken every night.   This had not a fufficient effect; therefore, on the 22d of *April*, I ordered the infufion prefcribed two years before, which foon removed his complaints.

He died foon afterwards, fairly worn out, in his ninetieth year.

C A S E   CVI.

November 2d.   Mr. S———, of B——h——, Æt. 61. Hydrothorax and fwelled legs. Squills were given for a week in very full dofes, and other modes of relief attempted; but his breathing became fo bad, his countenance fo livid, his pulfe fo feeble, and his extremities fo cold, that I was apprehenfive upon my fecond vifit that he had not twenty-four hours to live.   In this fituation I gave him the Infufum Digitalis ftronger than ufual, viz. two drams to eight ounces. Finding himfelf relieved by this, he continued to take it, contrary to the directions given, after the diuretic effects had appeared.

The

The ficknefs which followed was truly alarming; it continued at intervals for many days, his pulfe funk down to forty in a minute, every object appeared green to his eyes, and between the exertions of reaching he lay in a ftate approaching to fyncope. The ftrongeft cordials, volatiles, and repeated blifters barely fupported him. At length, however, he did begin to emerge out of the extreme danger into which his folly had plunged him; and by generous living and tonics, in about two months he came to enjoy a perfect ftate of health.

C A S E   CVII.

November 19th. Mafter S———, Æt. 8. Afcites and anafarca. A dram of Fol. Digitalis in a fix ounce infufion, given in dofes of a fpoonful, effected a perfect cure, without producing naufea.

1 7 8 3.

The reader will perhaps remark, that from the middle of *January* to the firft of *May*, not a fingle cafe occurs, and that the amount of cafes is likewife lefs than in the preceding or enfuing years; to prevent erroneous conjectures or conclufions, it may be expedient to mention, that the ill ftate of my own health obliged me to retire from bufinefs for fome time in the fpring of the year, and that I did not perfectly recover until the following fummer.

CASE

C A S E   CVIII.

January 15th. Mrs. G——, Æt. 57. A very fat woman; has been dropfical fince *November* laft; with fymptoms of difeafed vifcera. Various remedies having been taken without effect, an Infufion of Digitalis was directed twice a day, with a view to palliate the more urgent fymptoms. She took it four days without relief, and as her recovery feemed impoffible it was urged no farther.

C A S E   CIX.

May 1ft. Mrs. D————, Æt. 72. A thin woman, with very large anafarcous legs and thighs; no appetite and general debility. After a month's trial of cordials and diuretics of different kinds, the furgeon who had fcarified her legs apprehended they would mortify; fhe had very great pain in them, they were very red and black by places, and extremely tenfe. It was evident that unlefs the tenfion could be removed, gangrene muft foon enfue. I therefore gave her Infufum Digitalis, which increafed the fecretion of urine by the following evening, fo that the great tenfion began to abate, and together with it the pain and inflammation. She was fo feeble that I dared not to urge the medicine further, but fhe occafionally took it at intervals until the time of her death, which happened a few weeks afterwards.

CASE

C A S E   CX.

May 18th. I was defired to prefcribe for Mary Bowen, a poor girl at Hagley. Her difeafe appeared to me to be an ovarium dropfy. In other refpects fhe was in perfect health. I directed the Digitalis to be given, and gradually pufhed fo as to affect her very confiderably. It was done; but the patient ftill carries her big belly, and is otherwife very well.

C A S E   CXI.

May 25th. Mr. G——, Æt. 28. In the laft ftage of a pulmonary confumption of the fcrophulous kind, took an Infufion of Digitalis, but without any advantage.

C A S E   CXII.

May 31ft. Mr. H——, Æt. 27. In the laft ftage of a phthifis pulmonalis became dropfical. He took half a pint of the Infufum Digitalis in fix days, but without any fenfible effect.

C A S E   CXIII.

June 3d. Mafter B——, of D——, Æt. 6. With an univerfal anafarca, had an extremely troublefome cough. An opiate was given to quiet the cough at night, and 2 tea fpoonfuls of Infuf. Digit. were ordered every fix hours. The dropfy was prefently removed; but the cough continued, his
flefh

flefh wafted, his ftrength failed, and fome weeks afterwards he died tabid.

C A S E   CXIV.

June 19th. Mrs. L——, Æt. 28. A dropfy in
the laft ftage of a phthifis. Infufum Digitalis was
tried to no purpofe.

C A S E   CXV.

June 20th. Mrs. H——, Æt. 46. A very fat,
fhort woman; had fuffered feverely through the laft
winter and fpring from what had been called afthma;
but for fome time paft an univerfal anafarca pre-
vailed, and fhe had not lain down for feveral weeks.
After trying vitriolic acid, tincture of cantharides,
fquills, &c. without advantage, fhe took half a pint
of Infuf. Digitalis in three days.   In a week afterwards the dropfical fymptoms difappeared, her
breath became eafy, her appetite returned, and fhe
recovered perfect health.   The infufion neither
occafioned ficknefs nor purging.

C A S E   CXVI.

June 24th. Mrs. B——, Æt. 40. A puerperal
fever, and fwelled legs and thighs. The fever not
yielding to the ufual practice, I directed an Infufion
of Fol. Digitalis. It proved diuretic; the fwellings
fubfided, but the fever continued, and a few days
afterwards a diarrhœa coming on, fhe died.

CASE

C A S E   CXVII.

July 22d. Mr. F——, Æt. 48. A ftrong man, of a florid complexion, in confequence of intemperance became dropfical, with fymptoms of difeafed vifcera, great dyfpnœa, a very troublefome cough, and total lofs of appetite. He took mild mercurials, pills of foap, rhubarb, and tartar of vitriol, with foluble tartar and dulcified fpirits of nitre in barley water. After a reafonable trial of this plan, he took fquill every fix hours, and a folution of affafetida and gum ammoniac, to eafe his breathing: finding no relief, I gave him chryftals of tartar with ginger; but his remaining health and ftrength daily declined, and he was not at all benefited by the medicines. I was averfe to the ufe of Digitalis in this cafe, judging from what I had feen in fimilar inftances of tenfe fibre, that it would not act as a diuretic. I therefore once more directed fquill, with decoction of feneka and fal fodæ; but it was inefficacious. His ftrength being much broken down, I then ordered gum ammoniac, with fmall dofes of opium, and infufum amarum, continuing the fquill at intervals. At length I was urged to give the Digitalis, and confidering the cafe as defperate, I agreed to do it. The event was as I expected; no increafe in the urine took place; and the medicine being ftill continued, his pulfe became flow, and he apparently funk under its fedative effects. He was neither purged nor vomited; and had the Digitalis either been omitted alto-

altogether, or fufpended upon its firft effects upon
the pulfe being obferved, he might perhaps have
exifted a week longer.

C A S E   CXVIII.

July 26th.   Mr. W——, of W——, Æt. 47.
Phthifis pulmonalis, jaundice, afcites, and fwelled
legs. As it was probable that the only relief I could
give in a cafe fo circumftanced, would be by carry-
ing off the effufed fluids. I tried fquill and fixed
alkaly; and thefe failing, I ordered the Infufum
Digitalis. This had the defired effect, and, I believe, prolonged his life a few weeks.

C A S E   CXIX.

Auguft 15th.   Mrs. C———, Æt. 60. Afcites,
anafarca, difeafed vifcera, paucity of urine, and
total lofs of appetite. Thefe complaints had heretofore exifted repeatedly, and had been removed
by deobftruent and diuretic medicines; but in this
attack the fymptoms were fuffered to exift a longer
time and in a greater degree, before affiftance was
fought for. The remedies that ufed to relieve her
were now exhibited to no purpofe. Mild mercurials, foap, rhubarb, and fquill were tried; but fhe
grew rapidly worfe. Saline draughts with acetum
fcilliticum feemed for a few days to check the progrefs of her complaint, but they foon loft their effect, and diarrhœa enfued upon every attempt to
increafe the frequency of the dofe. Draughts with
Infuf. Digital. were then directed to be taken twice
a day.

a day. The effect was a powerful action on the kidneys, and a reduction of the swellings, but without sickness. A degree of appetite returned, but still the tendency to diarrhœa existed, and kept her weak. Tonic medicines were then tried, but without advantage, and in a month it was necessary to have recourse to the Digitalis again. It was directed in a half pint mixture; an ounce to be taken thrice in twenty-four hours. On the 2d day, finding her symptoms very much relieved, she took in the absence of her nurse, nearly a double dose of the medicine. The confequence was great sickness, languor continuing for several days, and almost a total stop to the fecretion of urine, from the time the sickness commenced.

The cafe now became totally unmanageable in my hands, and, after a fortnight, I was dismissed, and another physician called in; but she did not long survive.

This was not the first, nor the last instance, in which I have seen too large a dose of the medicine, defeat the very purpofe for which it was directed.

C A S E   CXX.

Auguft 22d. Mrs. S——, Æt. 36. Extreme faintinefs; anafarcous legs and thighs; great difficulty of breathing, troublesome cough, frequent chilly fits succeeded by hot ones; night fweats, and a tendency to diarrhœa. Apprehensive that the

more

more urgent fymptoms were caufed by water in the
lungs, I directed an Infufion of Digitalis, with an
ounce of diacodium to the half pint to prevent it
purging, a wine glafs full to be taken every night at
bed-time, and a mixture with confect. cardiac. and
pulv. ipecac. to be given in fmall dofes after every
loofe ftool.

On the fourth day fhe was better in all refpects;
had made a large quantity of water and did not purge.
In a few days more fhe loft all her complaints, except the cough, which gradually left her, without
any further affiftance.

I was agreeably deceived in the event of this cafe,
for I expected after the water was removed, to have
had a phthifis to contend with.

C A S E   CXXI.

Auguft 25th. T——W——, Efq; Æt, 50. A free
liver, difeafed vifcera, belly very tenfe, and much
fwollen; fluctuation perceptible, but the fwelling
circumfcribed; pulfe 132. This gentleman was under the care of my very worthy friend Dr. Afh,
who, having tried various modes of cure to no purpofe, afked me if I thought the Digitalis would
anfwer in this cafe. I replied that it would not,
for I had never feen it effectual where the fwelling
appeared very tenfe and circumfcribed. It was tried
however, but did not leffen the fwelling. I mention
this cafe, to introduce the above remark, and alfo
to

to point out the great effect the Digitalis has upon the action of the heart; for the pulse came down to 96. He was afterwards tapped, and continued, for some time under our joint attendance, but the pulse never became quicker, nor did the swelling return.

C A S E   CXXII.

September 7th. Mr. L——, Æt. 43. After several severe attacks of ill formed gout, attended for some time past with jaundice and other symptoms of diseased viscera, the consequences of intemperate living, was sent to Buxton ; from whence he returned in three weeks with ascites and anasarca. Under this complicated load of disease, I prescribed repeatedly without advantage, and at length gave him the Digitalis, which carried off the more obvious symptoms of dropsy ; but the jaundice, loss of appetite, diseased viscera, &c. rendered his recovery impossible.

1784.

C A S E   CXXIII.

February 12th. Mrs. C——, Æt. 54. A strong short woman of a florid complexion ; complained of great fullness across the region of the stomach; short breath, a troublesome cough, loss of appetite, paucity of urine ; and had a brownish yellow tinge on her skin and in her eyes. She dated these complaints from a fall she had through a trap door about the beginning of winter. From the beginning of January to this time, she had been repeatedly let

F   blood

blood, had taken calomel purges with jallap ; pills
of foap, rhubarb and calomel ; faline julep with
acet. fcillit. nitrous decoction, garlic, mercury rubbed
down. infus. amarum purg. &c.   After the failure
of medicines fo powerful, and feemingly fo well
adapted, and during the ufe of which all the fymptoms continued to increafe, it was evident that a
favourable event could not be expected.   However,
I tried the infufum Digitalis, but it did nothing.   I
then gave her pills of quickfilver, foap and fquill,
with decoction of dandelion, and after fome time,
chryftals of tartar with ginger.   Nothing fucceeded
to our wifhes, and the increafe of orthopnoea compelled me occafionally to relieve her by draftic
purges, but thefe diminifhed her ftrength, more in
proportion than they relieved her fymptoms. Tincture of cantharides, fal diureticus and various other
means were occafionally tried, but with very little
effect, and fhe died towards the end of March.

C A S E   CXXIV.

March 31ft.   Mifs W——, Æt. 60. Had been
fubject to peripneumonic affections in the winter.
She had now total lofs of appetite, very great debility, difficult breathing; much cough, a confiderable
degree of expectoration, and a paucity of urine. She
had been blooded, taken foap, affaf. and fquill,
afterwards affaf. and ammon. with acet. fcillit.
but all her complaints increafing, a blifter was applied to her back, and the Digitalis infufion directed
to be taken every night. The effect was an increafed

fecresecretion of urine, a confiderable relief to her breath, and fome return of appetite ; but foon afterwards fhe became hectic, fpat purulent matter, and died in a few weeks.

C A S E CXXV.

April 12th. Mrs. H——, of L——, Æt. 61. In *December* laft this Lady, then upon a vifit in London, was attacked with fevere fymptoms of peripneumony. She was treated as an afthmatic patient, but finding no relief, fhe made an effort to return to her home to die. In her way through this place, the latter end of December, I was defired to fee her. By repeated bleedings, blifters, and other ufual methods, fhe was fo far relieved, that fhe wifhed to remain under my care. After a while fhe began to fpit matter and became hectic. With great difficulty fhe was kept alive during the difcharge of the abfcefs, and about the end of March fhe had fwelled legs, and unequivocal fymptoms of dropfy in the cheft. Other diuretics failing, on the 12th of April I was induced to give her the Digitalis in fmall dofes. The relief was great and effectual. After an interval of fifteen days, fome fwellings ftill remaining in the legs, I repeated the medicine, and with fuch good effect, that fhe loft all her complaints, got a keen appetite, recovered her ftrength, and about the end of May undertook a journey of fifty miles to her own home, where fhe ftill remains in perfect health.

CASE

C A S E   CXXVI.

April 17th. Mr. F——, Æt. 59. A very fat
man, and a free liver ; had long been fubject to
what was called afthma, particularly in the winter.
For fome weeks paft his legs fwelled, he had great
fenfe of fullnefs acrofs his ftomach ; a fevere cough ;
total lofs of appetite, thirft great, urine fparing,
his breath fo difficult that he had not lain down in
bed for feveral nights. Calomel, gum ammoniac,
tincture of cantharides, &c. having been given in
vain, I ordered two grains of pulv. fol. Digitalis
made into pills, with aromatic fpecies and fyrup, to
be given every night. On the third day his urine
was lefs turbid ; on the fourth confiderably increafed in quantity, and in ten days more he was
free from all complaints, and has fince had no
relapfe.

C A S E   CXXVII.

May 7th. Mifs K——, Æt. 8. After a long
continued ague, became hectic and dropfical. Her
belly was very large, and fhe had a total lofs of appetite. Half a grain of fol. Digital. pulv. with 2
gr. of merc. alcalis. were ordered night and morning, and an infufion of bark and rhubarb with fteel
wine to be given in the day time. Her belly began
to fubfide in a few days, and fhe was foon reftored
to health. Two other children in the family,
affected nearly in the fame way, had died, from the
parents being perfuaded that an ague in the fpring
was

was healthful and fhould not be ftopped.—I know not how far the recovery in this cafe may be attributed to the Digitalis, but the child was fo near dying that I dared not truft to any lefs efficacious diuretic.

C A S E CXXVIII.

June 13th. Mr. C——, Æt. 45. A fat man, had formerly drank hard, but not latterly : laft March began to complain of difficult breathing, fwelled legs, full belly, but without fluctuation, great thirft, no appetite ; urine thick and foul ; complection brownifh yellow. Mercurial · medicines, diuretics of different kinds, and bitters, had been trying for the laft three months, but with little advantage. I directed two grains of the fol. Digital. in powder to be taken every night, and infuf. amar. with tinct. facr. twice a day. In three days the quantity of his urine increafed, in ten or twelve days all his fymptoms difappeared, and he has had no relapfe.

C A S E CXXIX.

June 17th. Mr. N——, of W——, Æt. 54. A large man, of a pale complexion ; had been fubject to fevere fits of afthma for fome years, but now worfe than ufual. The intermitting pulfe, the great difturbance from change of pofture, and the fwelled legs induced me to conclude that the exacerbation of his old complaint was occafioned by ferous effufion. I directed pills with a grain and half of the

F 3   pulv.

pulv. Digital. to be taken every night, and as he was
coftive, jallap made a part of the compofition.   He
was alfo directed to take muftardfeed every morning
and a folution of affafetida twice in the day.   The
effect of this plan was perfectly to our wifhes, and
in a fhort time he recovered his ufual health. About
half a year afterwards he died apoplectic.

C A S E   CXXX.

Mary B——. A young unmarried woman. Her
difeafe appeared to me a dropfy of the right ovarium.   She took an infufion of Digitalis, but, as I expected with no good effect.   She is ftill, I am
informed, nearly in the fame ftate.

C A S E   CXXXI.

July. 12th. Mrs. A——, of C———, Æt. 56.
After a feries of indifpofitions for feveral years,
became dropfical ; and had long been confined to
her chamber, unable to lie down or to walk.   She
was fo feeble, her legs fo much fwelled, her breath
fo fhort, and the fymptoms of difeafed vifcera fo
ftrong, that I dared not to entertain hopes of a cure ;
but wifhing to relieve her more urgent fymptoms,
directed quickfilver rubbed down and fol. Digital.
pulv. to be made into pills : the dofe, containing
two grains of the latter, to be given night and
morning.   She was alfo ordered to take a draught
with a dram of æther twice a day, and to have fca-
pulary iffues. Her breath was fo much relieved,
that

that fhe was able foon afterwards to come down
ftairs ; but her conftitution was too much broken to
admit of a recovery.

C A S E   CXXXII.

July 16th. Mr. B——, of W——, Æt. 31.
After a tertian ague of 12 months continuation, fuffered great indifpofition for 10 months more.   He
chiefly complained of great ftraitnefs and pain in
the hypochondriac region, very fhort breath,
fwelled legs, want of appetite.   He had been under
the care of fome very fenfible practitioners, but his
complaints increafed, and he determined to come to
Birmingham.   I found him fupported upright in
his chair, by pillows, every attempt to lean back
or ftoop forward giving him the fenfation of inftanta-
neous fuffocation.   He faid he had not been in bed
for many weeks.   His countenance was funk and
pale ; his lips livid ; his belly, thighs and legs
very greatly fwollen ; hands and feet cold, the
nails almoft black, pulfe 160 tremulous beats in a
minute, but the pulfation in the carolid arteries
was fuch as to be vifible to the eye, and to
fhake his head fo that he could not hold it ftill.
His thirft was very great, his urine fmall in quantity,
and he was difpofed to purge.   I immediately
ordered a fpoonful of the infufum Digitalis every
fix hours, with a fmall quantity of laudanum, to
prevent its running off by ftool, and decoction of
leontodon taraxacum to allay his thirft.   The next
day he began to make water freely, and could
allow

allow of being put into bed, but was raifed high
with pillows. Omit the infufion. That night he
parted with fix quarts of water, and the next night
could lie down and flept comfortably. *July* 21ſt.
he took a mild mercurial bolus. On the 25th. the
diuretic effects of the Digitalis having nearly ceafed,
he was ordered to take three grains of the pulv.
Digital. night and morning, for five days, and a
draught with half an ounce of vin. chalyb. twice a
day. *Auguſt* 15th. He took a purge of calomel and
jallap, and fome fwelling ſtill remaining in his legs,
the Digitalis infufion was repeated. The water
having been thus entirely evacuated, he was ordered faline draughts with acetum fcilliticum and
pills of falt of ſteel and extract of gentian. About
a month after this, he returned home perfectly well.

C A S E   CXXXIII.

July 28th. Mr. A—— of W——, Æt. 29, became dropfical towards the clofe of a pulmonary
confumption. He was ordered 12 grains of pulv.
fol. cicutæ and 1 of Digitalis twice a day. No remarkable effect took place.

C A S E   CXXXIV.

July 31. Mr. M——, Æt. 37. Hydrothorax.
A fingle grain of fol. Digital. pulv. taken every
night for three weeks cured him. The medicine
never made him fick, but increafed his urine, which
became clear ; whereas before it had been high coloured and turbid.

C A S E

C A S E   CXXXV

August 6th.   Mr. C—— of B————, Æt. 42.
Afthma and anafarca, the confequence of free living.   He had been for fome time under the care of
an eminent phyfician of this place, but his complaints proving unufually obftinate, he confulted
me.   I directed an infufion of Digitalis to be taken
every night, and a mixture with fquill and tincture
of cantharides twice every day.   In about a week
he became better, and continued daily mending.
He has fince enjoyed perfect health, having quitted
a line of bufinefs which expofed him to drink too
much.

C A S E   CXXXVI.

August 6th. Mr. M—— of C————, Æt. 44. Afcites
and anafarca, preceded by fymptoms of the epileptic
kind.   He was ordered to take two grains of pulv.
Digitalis every morning, and three every night;
likewife a faline draught with fyrup of fquills, every
day at noon.   His complaints foon yielded to this
treatment, but in the month of November following
he relapfed, and again afked my advice. The Digitalis alone was now prefcribed, which proved as efficacious as in the firft trial.   He then took bitters
twice a day, and vitriolic acid night and morning,
and now enjoys good health.

Before the Digitalis was prefcribed, he had taken
jallap purges, foluble tartar, falt of fteel, vitriol of
copper, &c.

C A S E

C A S E   CXXXVII.

Auguſt 10th. Mrs. W——, Æt. 55.   An anafarcous leg, and ſciatica ; full habit.   After bleeding and a purge, a bliſter was applied in the manner recommended by Cotunnius ; and two grains
of fol. Digital. with fifteen of fol. cicutæ were directed to be taken night and morning.   The medicine acted only as a diuretic ; the pain and ſwelling
of the limb gradually abated ; and I have not heard
of any return.

I muſt here bear witneſs to the efficacy of Cotunnius's method of bliſtering in the ſciatica, having
uſed it in a great number of caſes, and generally
with ſucceſs.

C A S E   CXXXVIII.

Auguſt 16th. Mrs. A—— of S——, Æt. 78.
About the middle of Summer began to complain of
ſhort breath, great debility, and loſs of appetite. At
this time there were evident marks of effuſion in the
thorax, and ſome ſwelling in the legs.   The advanced age, the weakneſs, and other circumſtances
of this patient, precluded every idea of her recovery;
but ſomething was to be attempted.   Squills and
other remedies had been tried ; I therefore directed
pills with two or three grains of the pulv. Digitalis
to be taken every night for ſix nights, and a ſaline
draught with forty drops of acetum ſcillit. twice in
the day. She took but few of the draughts, ſeldom
                                                              more

more than half one at a time, for they purged her, and she disliked them. The pills she took regularly, and with the happiest effect, for she could lie down, her breath was very much relieved, and a degree of appetite returned. *Sept.* 4th, some return of her symptoms demanded the further use of diuretics. I was afraid to push the Digitalis in so hazardous a subject, and therefore directed tinct. amara with tinct. canthar. and pills of squill, seneka, salt of tartar and gum ammoniac. These medicines did not at all check the progress of the disease, and on the 26th it became necessary to give the Digitalis again. The pills were therefore repeated as before, and infuf. amarum with fixed alkaly ordered to be taken twice a day. The event was as favorable as before ; and from this time she had no considerable return of dropsy, but languished under various nameless symptoms, until the middle or end of November.

C A S E CXXXIX.

Aug. 16th. Mrs. P—— of S——, Æt. 50. For a particular account of this patient, see Mr. Yonge's second Case.

C A S E CXL.

Sept. 20th. B—— B——, Esq. A true spasmodic asthma of many years continuance. After every method of relief had failed; both under my management, and also under the direction of several of the ablest physicians of this kingdom; I was induced to
give

give him an infusion of the Digitalis. It was continued until naufea came on, but procured no relief.

C A S E   CXLI.

October 5th. Mr. R——, Æt. 43. *(The patient
mentioned at No.* 102.*)* He had purfued his former
mode of life, and had now a return of his complaints, with evident marks of difeafed vifcera. His
belly not very large, but uncommonly tenfe. From
this circumftance I did not expect the Digitalis to
fucceed, and therefore tried for fome time to relieve him by the faline julep, with acet fcillitic.
jallap, mercury, fyrup of fquill, with aq. cinnam. decoction of Dandelion, &c.; but thefe being adminiftered without advantage, I was driven to the
Digitalis.   As he was very weak and much emaciated, I only gave two grains night and morning for
five days.   As no increafe of urine took place, I
ufed alkaline falt with tinct. cantharides :—This
proving equally unfuccefsful, on the 18th, I directed
two ounces of the infufum Digitalis night and morning. This was continued until naufea took place,
but the kidney fecretion was not increafed.   Squill
with opium, deobftruents of different kinds, fubli-
mate folution, fixed alkaly, tobacco infufion, were
now fuccefsively tried, but with the fame want of
fuccefs. The fullnefs of his belly made it neceffary
to tap him, and by repeating this operation he
continued alive to the end of the year.

C A S E

C A S E   CXLII.

October 19th. Mrs. R—— of B————, Æt. 47.
Suppofed Afthma, of eighteen months duration. She
had kept her room for four months, and could not
lie down without great difturbance; was very thin,
and had totally loft all inclination for food. She
was directed to take two gr. of pulv. fol. Digital.
night and morning for five days, and infufum amarum, at the hours of eleven and five. In the courfe
of a week fhe was much relieved, and could remain
in bed all night. After a few days interval fhe took
the Digitalis for five days more, and was foon after
that well enough to come down ftairs and conduct
her family affairs.

In *April* 1785, fhe had a flight return, but not
fuch as to confine her to her chamber. She experienced the fame relief from the fame medicine, but
continuing it for feven days without interruption, it
excited naufea.

C A S E   CXLIII.

October 28th. Mr. A——, fubject to nephritis
calculofa : After an attack of that kind, had ftill a
troublefome fenfe of weight about his loins, now and
then rifing to pain, and a degree of dyfuria, together with a want of appetite. Thefe fymptoms not
readily yielding to the ufual methods of treatment,
I directed an infufion of Digitalis. The fourth dofe
caufed

caufed a copious flow of urine ; the fixth made him fick, and he was more or lefs fick at times for three days ; but felt no more of his complaints.

I don't believe it is at all neceffary to bring on ficknefs in thefe cafes, but an unexpected abfence from town prevented me from feeing him time enough to ftop the exhibition of the medicine.

C A S E   CXLIV.

October 31ft. Mrs. C——, of W——, Æt. 67. Afthma, and very thick hard legs of long continuance. The laft month or two her breath worfe than ufual, her belly fwollen, her thighs anafarcous, and her urine in fmall quantity. After trying garlic, fquill, and purgatives without advantage, I directed the Digital. Infuf. After taking about five ounces, her urine from thick and turbid, changed to clear and amber coloured, its quantity confiderably increafed, and her breathing eafy. Contrary to my orders, but impelled by the relief fhe had found, fhe finifhed the remaining three ounces of the infufion, which made her very fick, and the free flow of urine immediately ceafed. No medicine was adminiftered for a fortnight, during which time her complaints increafed. I then directed an infufion of tobacco, which affected her head, but did not increafe her urine. She had recourfe again to the Digitalis infufion, which once more removed the fulnefs of the belly, reduced the fwellings of her thighs, and relieved her breath, but had no effect upon her legs.

C A S E

C A S E   CXLV.

Nov. 2d. Miſs B—— of C——, Æt. 22. A very evident fluctuation in the abdomen, which was conſiderably diſtended, whilſt the reſt of her frame was greatly emaciated. The preſence of cough, hectic fever, and other circumſtances, made it probable that this apparent aſcites was cauſed by a purulent, and not a watery effuſion. However it was poſſible I might be miſtaken; the Digitalis was therefore given, but without any advantage.

The further progreſs of the diſeaſe confirmed my firſt opinion, and ſhe died conſumptive.

C A S E   CXLVI.

Nov. 4th. Mr. P—— of M——, Æt. 40. Subject to troubleſome nephritic complaints, and after the laſt attack did not recover, or void the gravelly concretions as uſual, a ſenſe of weight acroſs his loins continuing very troubleſome. The uſual medicines failing to relieve him, I ordered four grains of pulv. Digital. to be taken every other night for a week, and fifteen grains of mild fixed vegetable alkaly to be ſwallowed twice a day in barley water. He ſoon loſt all his complaints; but we muſt not in this caſe too haſtily attribute the cure to the Digitalis, as the alkaly has alſo been found a very uſeful medicine in ſimilar diſorders.

C A S E   CXLVII.

Nov. 4th. Mr. B—— of N——, Æt. 60. Had been much fubject to gout, but his conftitution being at length unable to form regular fits, he became dropfical. Pulv. fol. Digital. in dofes of two or three grains, at bed-time, gave him fome relief, but did not perfectly empty him. About three months afterwards he had occafion to take it again ; but it then produced no effect, and he was fo debilitated that it was not urged further.

C A S E   CXLVIII.

Nov. 8th. Mr. G—— Æt. 35. In the laft ftage of a phthifis pulmonalis, was attacked with a moft urgent and painful difficulty of breathing. Sufpecting this diftrefs might arife from watery effufion in the cheft, I gave him Digitalis, which relieved him confiderably ; and during the remainder of his life his breath never became fo bad again.

C A S E   CXLIX.

Nov. 13th. Mrs. A—— of W——h——, Æt. 68. One of thofe rare cafes in which no urine is fecreted. It proved as refractory as ufual to remedies, and not having ever fucceeded in the cure of this difeafe, I determined to try the Digitalis. It was given in infufion, and, after a few dofes, the fecretion of a fmall quantity of urine feemed to juftify the attempt. The next day, however, the fecretion

cretion ceafed, nor could it be excited again, tho'
at laft the medicine was pufhed fo as to occafion
ficknefs, which continued at intervals for three
days.

C A S E   CL.

Nov. 20th. Mrs. B——, Æt. 28. In the laft
ftage of a pulmonary confumption became dropfical. I directed three grains of the pulv. Digital. to
be taken daily, one in the morning, and two at
night. She took twenty grains without any fenfible effect.

C A S E   CLI.

Nov. 23d. Mafter W——, Æt. 7. Suppofed
hydrocephalus internus. A grain of pulv. fol. Digitalis was directed night and morning. After
three days, no fenfible effects taking place, it was
omitted, and the mercurial plan of treatment
adopted. The child lived near five months afterwards. Upon diffection near four ounces of water
were found in the ventricles of the brain.

C A S E   CLII.

Nov. 26th. Mrs. W——, Æt. 65. I had attended this lady laft winter in a very fevere peripneumony, from which fhe narrowly efcaped with
her life. When the cold feafon advanced this winter, fhe perceived a difficulty in breathing, which gradually became more and more troublefome. I found

G   her

her much harraffed by a cough, which occafioned her
to expectorate a little: the leaft motion increafed her
dyfpnœa; fhe could not lie down in bed; her legs
were confiderably fwelled, her urine fmall in quantity.   I directed two grains of pulv. Digitalis made
into a pill with gum ammoniac, to be taken every
night, and to promote expectoration, a fquill mixture twice in the day.   Her urine in five days became clear and copious, and in a fortnight more fhe
loft all her complaints, except a cough, for which
fhe took the lac ammoniacum.

It is not improbable that the fquill might have
fome fhare in this cure.

C A S E   CLIII.

December 7th. Mr. H——, Æt. 42.   A large
fat man, very fubject to gravelly complaints.   After an attack in the ufual manner, continued to feel
numbnefs in his lower limbs, and a fenfe of weight
acrofs his loins.   I directed infufum Digitalis to be
given every fix hours.   Six ounces made him fick,
and he took no more.   The next day his urine increafed, a good deal of fand paffed with it, and he
loft his difagreeable feels, but the ficknefs did not
entirely ceafe before the fourth day from its com-
mencement.

CASE

C A S E CLIV.

December 27th. Mr. B——, of H——, Æt. 55.
Symptoms of hydrothorax, at firſt obſcurely, afterwards more diſtinctly marked. Many things were
tried, but the ſquill alone gave relief. At length
this failed. About the third month of the diſeaſe,
a grain of pulv. Digital. was ordered to be taken
night and morning. This produced the happieſt
effects. In *March* following he had ſome ſlight
ſymptoms of relapſe, which were ſoon removed by
the ſame medicine, and he now enjoys good health.
For a more particular narrative ſee caſe the firſt,
communicated by Mr. Yonge.

C A S E CLV.

December 31ſt. Mrs. B——, of E——, Æt. 50.
An ovarium dropſy of long continuance. She took
three grains of pulv. Digital. every night at bed
time, for a fortnight, but without any effect.

C A S E CLVI.

A poor man in this town, after his kidneys had
ceaſed to ſecrete urine for ſeveral days, was ſeized
with hickup, fits of vomiting, and tranſient delirium.
After examination I was ſatisfied the diſeaſe was
the ſame as that mentioned at CXLIX. A very experienced apothecary having tried various methods to
relieve him, I deſpaired of any ſucceſs, but determined to try the Digitalis. It was accordingly given

in infufion. At firft it checked the vomitings, but did not occafion any fecretion of urine.

1785.

The cafes which have occurred to me in the courfe of this year, are numerous ; but as the events of fome of them are not yet fufficiently afcertained, I think it better to withhold them at prefent.

HOSPITAL

HOSPITAL CASES,

Under the Direction of the Author.

THE four following cafes were drawn out at my requeſt by Mr. Cha. Hinchley, late apothecary to the Birmingham Hoſpital. They are all the Hoſpital cafes for which the Digitalis was preſcribed by me, whilſt he continued in that office.

C A S E   CLVII.

March 15th, 1780. John Butler, Æt. 30. Aſthma and ſwelled legs. He was directed to take myrrh and ſteel every day, and three ſpoonfuls of infuſum Digitalis every night. On the 8th of April he was diſcharged, cured of the ſwellings and ſomething relieved of his aſthmatic affections.

C A S E   CLVIII.

November 18th, 1780. Henry Warren, Æt. 60. This man had a general anaſarca and aſcites, and was moreover ſo aſthmatic, that, neither being able to ſit in a chair nor lie in bed, he was obliged conſtantly to walk about, or to lean forward againſt a window or table. You preſcribed for him thus.

G 3   R. Aq.

R. Aq. cinn. fpt. ʒiv.
    Oxymel. fcillit.
    Syr. fcillit. aa. ʒi.. m ⸴ap. cochlear. larg. fexta
    quaque horâ.

This medicine producing no increafed difcharge
of urine, on the 25th you ordered the infufion of
Digitalis, two fpoonfuls every four hours. After
taking this for thirty fix hours, his urine was difcharged in very great quantity ; his breath became
eafy, and the fwellings difappeared in a few days,
though he took no more of the medicine. On the
2d of *December* he was ordered myrrh and lac ammoniacum, which he continued until the 23d, when
he was difcharged cured, and is now in good health.

C A S E   CLIX.

November 3d, 1781. Mary Crockett, Æt. 40.
Afcites and univerfal anafarca. For one week fhe
took fal. diureticus and tincture of cantharides, but
without advantage. On the 10th you directed the
infufion of Digitalis, a dram and half to half a pint,
an ounce to be taken every fourth hour. Before
this quantity was quite finifhed, the urine began to
be difcharged very copioufly. The medicine was
then ftopped as you had directed. On the 15th,
being coftive, fhe took a jallap purge, and on the
24th fhe was difcharged cured.

C A S E   CLX.

March 16th, 1782. Mary Bird, Æt. 61. Great
fullnefs about the ftomach ; difeafed liver, and ana-
                                                    farcous

farcous legs and thighs.   For the firſt week ſquill
was tried in more forms than one, but without advantage.   On the 22d ſhe began with the Digitalis,
which preſently removed all the ſwelling.

She was then put upon the uſe of aperient medicines and tonics, and on the firſt of *Auguſt* was diſcharged perfectly cured.

———————

The three following Caſes were drawn up and com-
municated to me by Mr. Bayley, who ſucceeded
Mr. Hinchley as apothecary to the Hoſpital at
Birmingham:

DEAR SIR,   Shiffnal, April 26th, 1785.
          DURING my reſidence in the
Birmingham General Hoſpital, I had frequent op-
portunities of ſeeing the great effects of the Digitalis
in dropſy.   As the exhibition of it was in the following inſtances immediately under your own direction, I have drawn them up for your inſpection,
previous to your publiſhing upon that excellent
diuretic.   Of its efficacy in dropſy I have conſiderable evidence in my poſſeſſion, but conſider myſelf not at liberty to ſend you any other caſes except thoſe you had yourſelf the conduct of.   The
Digitalis is a very valuable acquiſition to medicine;
and, I truſt, it will ceaſe to be dreaded when it is
well underſtood.
          I am, Sir, your obedient,
               And very humble ſervant,
                    W. BAYLEY.
                         CASE

C A S E   CLXI.

Mary Hollis, aged 62, was admitted an out patient of the Birmingham General Hofpital *February*
12th, 1784, labouring under all the effects of hydrothorax; her dread of fuffocation during fleep
was fo great, that fhe always repofed in an elbow
chair. She was directed to take two grains of Digitalis in powder every night and morning, and for
a few days found great relief; but, on the eighth
day, as fhe had complained of ficknefs, and had
been confiderably purged, fhe was ordered to defift
taking any more of her powders. On the 14th day
fhe was ordered an ounce of the following infufion
twice in a day: R. Fol. Digital. purp. ficc. 3ifs. aq.
bullient. ℔fs. digere per femi-horam, colaturæ adde
tinct. aromatic 3i. This infufion did not purge,
but fometimes excited naufea, though not fufficient
to prevent her from continuing its ufe. She grew
gradually better, and on the 6th of *May* was difcharged perfectly cured. The diuretic effects of the
Digitalis were in this inftance immediate.

C A S E   CLXII.

Edward James, Æt. 21. Admitted *March* 20th,
1784. Complained of great difficulty of breathing, pain in his head, and tightnefs about the ftomach, with a trifling fwelling of his legs. Ordered
pil. fcillit. Ʒi. ter de die. On the third day his legs
much more fwelled, his breathing more difficult,
and in every refpect worfe; his pulfe very fmall

and

and quick, complained when he turned in bed, of ſomething like water rolling from one ſide of the thorax to the other. A remarkable blueneſs about the mouth and eyes, and purged conſiderably from the pil. ſcill. Ordered to omit the pills and to take ʒi. of infuſ. Digitalis every eight hours; the proportion ʒiſs. to eight ounces of water and ʒi. of aq. n. m. ſp.—7th Day, The infuſion had neither purged, nor vomited him: he only complained once or twice of giddineſs. His belly was now very hard, rather black on the right ſide the navel, and his legs amazingly ſwelled. Ordered a bolus with rhubarb and calomel, to be taken in the morning, and ʒii. julep ſalin. cum tinct. canthar. gutt. forty ter die. —12th Day, nearly in the ſame ſtate, except his breathing which was ſomewhat more difficult, being now obliged to have his head conſiderably raiſed. Perſiſtat—From this day to the 32d day he became hourly worſe. His belly which at firſt was only hard, now evidently contained a large quantity of water, his legs were more ſwelled, and a large ſphacelated ſore appeared upon each outer ancle. Reſpiration was ſo much obſtructed, that he was obliged to ſit quite upright to prevent ſuffocation. He made very little water, not more than eight ounces in a day and a night, and was much emaciated. Ordered his purging bolus again, and ʒii. of a mixture with ſal diuretic. ʒſs. to ʒxii. three times in a day, and a poultice with ale grounds to his legs.

54th day. To this period there was not the leaſt probability of his exiſting ; his legs and thighs were
one

one continued blubber, his thorax quite flat, and his belly fo large that it meafured within one inch as much as a woman's in this Hofpital the day fhe was tapped, and from whom twenty feven pounds of coagulable lymph were taken. He made about three ounces of water in twenty-four hours: his penis and fcrotum were aftonifhingly fwelled, and no difcharge from the fores upon his legs. Ordered to take a pill with two grains of powdered Fox-glove night and morning. For a few days no fenfible effect, but about the 60th day he complained of being continually giddy, and had fome little pain in his ftomach. He now made much more water, and dared to fleep. His appetite which through the whole of his illnefs had been very bad, was alfo better.   66th day. Breathing very much relieved, the quantity of water he made was three chamber pots full in a day and a night, each pot containing two quarts and four ounces, moderately full. Ordered to continue his pills, and his legs which were very flabby, to be rolled.

69th day. His belly nearly reduced to its natural fize, ftill made a prodigious quantity of water, his appetite very good, habit of body rather lax, and his complexion ruddy. On the 2d of *June*, being ftill rather weak, he was ordered decoct. cort. ʒii. ter de die ; and on the 12th was difcharged from this Hofpital perfectly cured.

W. BAYLEY.

Mr.

Mr. Bayley's refpectful compliments to Doctor Withering: he fends the cafe of Edward James, which he believes is pretty correct. He laments not having it in his power to fend the meafure of his belly, having unfortunately miflaid the tape: he heard from James yefterday, and he is perfectly well.

General Hofpital, Auguft 5, 1784.

C A S E   CLXIII.

On the 26th *February*, 1785, Sarah Ford, aged 42, was admitted an out-patient of the Birmingham General Hofpital: fhe complained of confiderable pain in her cheft, and great difficulty of breathing, her face was much fwelled and her thighs and legs were anafarcous. She had extreme difficulty in making water, and with many painful efforts fhe did not void more than fix ounces in twenty-four hours. She had been in this fituation about fix weeks, during which time fhe had taken ammoniacum, olibanum, and large quantities of fquills, without any other effect than frequent ficknefs. Upon her commencing an Hofpital patient, the following medicine was exhibited. R. gum ammoniac ʒii. pulv. fol. Digital. purp. ꝺii. fp. lavand. comp. ut fiat pil. 40. cap. ii. nocte maneque. She continued the ufe of thefe pills for a few days, without any fenfible effect. On the eighth day her breathing was much relieved, her legs and thighs were not fo much fwelled, and in a day and
a night

a night she made five pints of water. By the 12th day her legs and thighs were nearly reduced to their natural size. She continued to make water in large quantities, and had lost her pain in the thorax. To the 20th of *March*, she made rapid advances towards health, when not a symptom of disease remaining, she was discharged.

COMMU-

COMMUNICATIONS

FROM CORRESPONDENTS.

London, Norfolk-ſtreet,
May 31ſt, 1785.

Sir,

I HAD the favour of your letter laſt week; and I ſhall be very happy if I can give you any intelligence relating to the Foxglove, that can anſwer the purpoſe in which you are ſo laudably engaged.

It is true that my brother, the late Dr. Cawley, was greatly relieved, and his life, perhaps, prolonged for a year, by a decoction of the Foxglove root; but why it had not a more laſting effect, it is neceſſary I ſhould tell you that he had all the ſigns of a diſtempered viſcera, long before any watery ſwellings appeared; it was manifeſt that his dropſy was merely ſymptomatic, and he could therefore only from time to time have any relief from medicine. In the year 1776, he returned from London to Oxon. having conſulted ſeveral phyſicians at the former place, and Dr. Vivian at the latter, but without any ſucceſs; and he was then told of a carpenter at Oxon. that had been cured of a Hydrops pectoris by the Foxglove root, and as he

was

was a younger, and in other refpects an healthy man, his cure, I believe, remains a perfect one.

I did not attend my brother whilft he took the medicine, and therefore I cannot fpeak precifely to the operation of it; but I remember, by his letters, that he was dreadfully fick and ill for feveral days before the fecretion of urine came on, but which it did do to a great degree; relieved his breath, and greatly leffened the fwelling in his legs and thighs; but the two inftances I have lately feen in this part of the world, are much ftronger proofs of the efficacy of it than my brother's cafe.

I am, &c.
ROBERT CAWLEY.

N. B. Whenever I have another opportunity of giving the Foxglove, it fhall be in fmall dofes:—In which I fhould hope it might fuceed, although it might be more flowly. If you fhould try it with fuccefs, I fhould be glad to know what mode you made ufe of.

Dr. Cawley's prefcription.

R. Rad. Digital..purpur. ficcat. et contuf. ℥ii.
    Coque ex aq. font. ℔ii. ad ℔i. colat. liquor. adde aq. junip. comp. ℥ii.
    Mell. anglic ℥i. m. fumat cochl. iv. omni nocte h. f. et mane.

I have

—I have elfewhere remarked, that when the Digitalis has been properly given, and the diuretic ef-effects produced, that an accidental over-dofe bringing on ficknefs, has ftopped the fecretion of urine. In the prefent inftance it likewife appears, that violent ficknefs may be excited, and continue for feveral days without being accompanied by a flow of urine; and it is probable that the latter circumftance did not take place, until the feverity of the former abated. If Dr. Cawley had not had a conftitution very retentive of life, I think he muft have died from the enormous dofes he took; and he probably would have died previous to the augmentation of the urinary difcharge. For if the root from which his medicine was prepared, was gathered in its active ftate, he did not take at each dofe lefs than *twelve* times the quantity a ftrong man ought to have taken. Shall we wonder then that patients refufe to repeat fuch a medicine, and that practitioners tremble to prefcribe it? Were any of the active and powerful medicines in daily ufe to be given in dofes *twelve* times greater than they are, and thefe dofes to be repeated without atttention to the effects, would not the patients die, and the medicines be condemned as dangerous and deleterious?—Yet fuch has been the fate of Foxglove!

A Letter

A Letter to the Author, from Mr. BODEN, Surgeon, at Brofeley, in Shropfhire.

Dear Sir,   Brofeley, 25th May, 1785.

HAVE inclofed the prefcriptions that contained the fol. Digital. which I gave to Thomas Cooke and Thomas Roberts.

Thomas Cooke, Æt. 49, had been ill about two or three weeks. When I faw him he had no appetite, and a conftant thirft: a fullnefs and load in the ftomach: the thighs, legs and hands, much fwell'd, and the face and throat in a morning; was coftive, and made but little water, which was high coloured; the pulfe very weak, and his breath exceeding bad. *June* 17th. R. Argent. viv ʒi. conf. cynofbat. ∋ii. fol. Digital. pulv. gr. xv. f. pil. xxiv. capt. ii. omni noᵉte horâ decubitus. He was likewife purged by a bolus of argent. viv. jallap, Digit. elaterium and calomel, which was repeated on the fourth day, to the third time. From *June* 17th to the 29th, the fymptoms were moftly removed, making water freely, and having plenty of ftools; in a week after he was perfeᵉtly well, and remains fo ever fince. The cure was finifhed by fteel and bitters.

Thomas Roberts, Æt. 40, had a deformed cheft, was obliged to be almoft in an ereᵉt pofture when in bed; the other fymptoms were nearly the fame as Cookc's. *Auguft* 3d. The pills prefcribed *June*
17th

17th for Cooke.——17th. A purging bolus of jalap and Digitalis, once a week. He continued the medicines till the latter end of *Auguſt*, when he got very well; but the complaint returned in *Jan.* worſe than before. He is now much better, but I have great reaſon to believe the liver to be diſeaſed.

I am, with the greateſt reſpect,

Your very obliged humble ſervant,

DANIEL BODEN.

P. S. The ſecond patient, on his relapſe, took Digitalis again, combined with other things.

CASE communicated by Mr. CAUSER, Surgeon, at Stourbridge, Worceſterſhire.

Mr. P—— of H—— M——, in the pariſh of Kingſwinford, aged about 60; had been a ſtrong healthy, robuſt, corpulent man; worked hard early in life at edge-tool making, and drank freely of ſtrong malt liquor; for many years had been ſubject to gout in the extremities; for a few years paſt had been very aſthmatic, and the gout in the extremities gradually decreaſed. When I firſt ſaw him, which was *Sept.* 12, 1779, his legs were anaſarcous, his belly much ſwelled, and an evident fluctuation of water. His breathing very bad, an irregular pulſe, and unable to lie down. His eaſieſt

H   poſture

posture was standing with his body leaning over a
chair, in which situation he would continue many
hours together, labouring for breath, with the sweat
trickling down his face very profusely; the urine
in very small quantity. Diuretics of every kind I
could think of were used with very little or no advantage. Blisters applied to the legs relieved very
considerably for a time, but by no means could I
increase the urinary discharge. Warm stomachic
medicines were given, and at the same time sina-
pisms applied to the feet, in hopes of enticing gout
to the extremities, but without any good effect.—
November 22d. The swelling considerably increasing,
an emetic of acet. scillitic. was given, which acted
very violently, and increased the urinary discharge
considerably. He continued better and worse, using
different kinds of diuretic and expectorating medicines until *September* 1781, when the disease was so
much worse, I did not expect he could live many
days. The acet. scillitic. was repeated, a table
spoonful every half hour, till it acted briskly upwards and downwards; but without increasing the
urinary discharge.—On the 17th of *September* I in-
fused ʒiii. of the fol. Digitalis in ℥vi. of boiling
water, for four hours; then strained it, and added
℥i. of tinct. aromatica.—On the 18th he began by
taking one spoonful, which he was to repeat every
half hour, till it made him very sick, unless giddi-
ness, loss of sight, or any other disgreeable effect
took place. I had never given the medicine before,
and had prepared him to expect the operation to
be very severe. I saw him again on the 21st; he
had

had taken the medicine regularly, till the whole quantity was confumed, without perceiving the leaft effect of any kind from it, and continued well till the evening of the following day, when a little ficknefs took place, which increafed, but never fo as to occafion either vomiting or purging; but a furprifing difcharge of urine. The faliva increafed fo as to run out of his mouth, and a watery difcharge from his eyes; thefe difcharges continued, with a continual ficknefs, till the fwelling was totally gone, which happened in three or four days. He afterwards took fteel and bitters; and continued very comfortably, without any return of his dropfy, until the the 7th of *April* 1782, when he was feized with an epidemic cough, which was very frequent with us at that time. His fwellings now returned very rapidly, with the greateft difficulty in breathing, and he died in a few days. Blifters and expectorating medicines were ufed on this laft return.

Extract of a Letter from Mr. CAUSER.

Mrs. S——, the fubject of the following Cafe, was as ill as it is poffible for woman to be and recover; from the inefficacy of the medicines ufed, I am convinced no medicine would have faved her but the Digitalis. I never faw fo bad a cafe recovered; and it fhews, that in the moft reduced ftate of body, the medicine in fmall dofes, will prove fafe and efficacious.

N. B. The

N. B. The Digitalis, in pills, never occafioned the leaft ficknefs.   She took two boxes of them.

C A S E.

January 2d, 1785.   Mrs. S——, of W——, near Kidderminfter, aged 38, has been affected with dropfical fwellings of her legs and thighs, about fix weeks, which have gradually grown worfe; has now great difficulty in breathing, which is much increafed on moving; a very irregular, intermittent pulfe, urine in very fmall quantity, and in the feventh month of her pregnancy: a woman of very delicate conftitution, with tender lungs from her infancy, and very fubject to long continued coughs.

R. Pulv. fcillæ gr. iii.
Jalap gr. x. fyr. rofar. folut. tinct. fenn. aa 3ii. aq. menth. v. fimpl. ℥ifs. m. mane fumend.

R. pulv. fcillæ ℈i. G. ammoniac. fapon. venet. aa 3ifs. fyr. q. f. f. pilul. 42 cap. iii. nocte maneque.

On the 7th found her worfe, and the fwelling increafed; the urine about ℥x in the twenty-four hours.

R. Fol. ficcat. Digital. 3iii. coque in. aq. fontan. ℥xii. ad ℥vi. cola et adde. aq. juniper. comp. ℥ii. facchar. alb. ℥fs. m. cap. cochlear. i. larg. 4tis horis.

She

She took about three parts of the medicine before any effect took place. The firſt was ſickneſs, ſucceeded by a conſiderable diſcharge of urine. She continued the medicine till the whole was conſumed, which cauſed a good deal of ſickneſs for three or four days.

I ſaw her again on the 12th. The quantity of urine was much increaſed, and the ſwelling diminiſhed. Pulſe and breathing better.

R. Fol. ſicc. Digital. G. aſſafetid. aa ʒi. calomel. pp. gr. x. ſp. lavand. comp. q. ſ. fiat pilul. xxxii. cap. ii. omni noƈte horâ ſomni.

A plentiful diſcharge of urine attended the uſe of theſe pills, and ſhe got perfeƈtly free from her dropſical complaints.

March 15th ſhe was delivered: had a good labour, was treated as is uſual, except in not having her breaſts drawn, not intending ſhe ſhould ſuckle her child, being in ſo reduced a ſtate. Continued going on well till the 18th, when ſhe was ſeized with very violent pains acroſs her loins, at times ſo violent as to make her cry out as much as labour pains. Enema cathartic. Fot. papav. applied to the part.

R. Pulv. ipecacoan. gr. vi. opii. gr. iv. ſyr. q. ſ. fiat pilul. vi. capt. i. 2da quaque horâ durante dolore.

H 3   R. Julep.

R. Julep. e camphor. fp. minder. aa ʒii. capt.
cochlear. i. larg. poft fingul. pilul.

19th. Breathing fhort, unable to lie down, very
irregular low pulfe fcarcely to be felt, fainty, and
a univerfal cold fweat: no appetite nor thirft, fpafmodic pains at times acrofs the loins very violent,
but not fo frequent as on the preceding day.

R. Gum ammoniac. affafetid. aa ʒi. camphor.
gr. xii. fiat pilul. 24. capt. ii. 3tia quaque
horâ in cochlear. ii. mixtur. feq.

R. Balfam. peruv. ʒiii. mucilag. G. arab. q. f.
flor. zinci g. vi. aq. menth. fimp. ℔fs. m.

Applic. Emp. veficat. femorib. internis.

R. Sp. vol. fœtid. elixir. paregor. balfam.
Traumatic. aa ʒiii. capt. cochlear. parv. ur-
gente languore.

20th. Much the fame; makes very little water,
and the legs begin to fwell.—Applic. Emp. e pice
burgund. lumbis.

23d. The fwelling very much increafed.—Capt.
gutt. xv. acet. fcillitic. ter die in two fpoonfuls of
the following mixture.

R. Infuf. baccar. juniper. ʒvi. tinct. amar. tinct.
ftomachic. aa ʒi. m.

25th.

25th. Much the fame.

28th. The fwelling confiderably increafed, in other refpects very much the fame.

30th. Breathing very bad, with cough and pain acrofs the fternum, unable to lie down, legs, thighs, and body very much fwelled, urine not more than four or five ounces in the twenty-four hours; hot and feverifh, with thirft.

Applic. Emp. veficat. ftomacho et fterno.

R. G. affafetid. ∂ii. pulv. jacob. ∂i. rad. fcill. recent. gr. xii. extract. thebaic. gr. iv. f. pilul. xvi. cap. iv. omni nocte.

R. Sal. nitr. fal. diuretic. aa ʒii. pulv. e contrayerv. comp. ʒi. facchar. ℥i. emulf. commun. ℔i. aq. cinnam. fimpl. ℥i. m. capt. cochlear. iv. ter die.

April 2d. Much the fame, no increafe of urine.

3d. Breathing much relieved by the blifter, which runs profufely. Repeated the medicines, and continued them till the

12th. The cough very bad, pulfe irregular, fwelling much increafed, urine in very fmall quantity, not at all increafed; great lownefs and fainting. She defired to have fome of the pills which relieved her

her fo much when with child. I was almoft afraid
to give them, but the inefficacy of the other medicines gave me no hopes of a cure from continuing them, which made me venture to comply with
her requeft.

R. Fol. ficcat. Digital. G. affafetid. aa ʒi. fp. lavand. comp. q. f. f. pilul. xxxii. cap. ii. omni mane; et omni noɕe cap. pilul. e ftyrace
gr. vi.

17th. Confiderable increafe of urine.

21ft. Swelling a good deal diminifhed; urine
near four pints in twenty-four hours, which is more
than double the quantity fhe drinks.

Applic. Emp. veficat. femoribus internis.

The Digitalis pills and opiate at bed-time continued. Takes a tea cup of cold chamomile tea every morning.

25th. Swelling much diminifhed, makes plenty
of water, appetite much mended, cough and breathing better. She omitted the medicine for three
days; the urine began to diminifh, the fwelling
and fhortnefs of breathing worfe. On repeating it
for two days, the difcharge was again augmented,
and a diminution of the fwelling fucceeded. She
has continued the pills ever fince till the 14th of
May;

May; the dropfical fymptoms and cough are entirely gone, the water is in fufficient quantity, her
ftrength is recovered, and fhe has a good appetite.
All fhe now complains of is a weight acrofs her ftomach, which is worfe at times, and fhe thinks, unlefs it can be removed, fhe fhall have a return of
her dropfy.

Extract of a Letter from Doctor FOWLER, Phyfician, at Stafford.

I UNDERSTAND you are going to publifh
on the Digitalis, which I am glad to hear, for I have
long wifhed to fee your ideas in print about it, and
I know of no one (from the great attention you
have paid to the fubject) qualified to treat on it but
yourfelf.   There are gentlemen of the faculty who
give verbal directions to poor patients, for the pre-
paring and taking of an infufion or decoction of the
green plant. Would one fuppofe that fuch gentlemen had ever attended to the nature and operation
of a fedative power on the functions, *particularly*
the *vital?* Is not fuch a vague and unfcientific mode
of proceeding putting a two edged fword into the
the hands of the ignorant, and the moft likely method to damn the reputation of any very active and
powerful medicine? And is it not more than probable
that the *neglect* of adhering to a *certain* and *regular*
preparation of the nicotiana, and the *want* (of what
you *emphatically* call) a *practicable* dofe, have been
the chief caufes of the once rifing reputation of
                                                    that

that noted plant being damned above a century ago? In fhort, the Digitalis is beginning to be ufed in dropfies, (although fome patients are faid to go off fuddenly under its adminiftration) fomewhat in the ftyle of broom afhes; and, in my humble opinion, the public, at this very inftant, ftand in great need of your *precepts*, *guards*, and *cautions* towards the fafe and fuccefsful ufe of fuch a powerful fedative diuretic; and I have no doubt of your minute attention to thofe particulars, from a regard to the good and welfare of mankind, as well as to your own reputation with refpect to that medicine.

I remember an officer in the Staffordfhire militia, who died here of a dropfy five years ago. The Digitalis relieved him a number of times in a wonderful manner, fo that in all probability he might have obtained a radical cure, if he would have refrained from hard drinking. I underftood it was firft ordered for him by a medical gentleman, and its fedative effects proved fo mild, and diuretic operation fo powerful, that he ufed to prepare it afterwards for himfelf, and would take it with as little ceremony as he would his tea. It is faid, that he was fo certain of its fuccefsful operation, that he would boaft to his bacchanalian companions, when much fwelled, you fhall fee me in two days time quite another man.

CASES

CASES communicated by Mr. J. Freer, jun. Surgeon, in Birmingham.

C A S E   I.

Nov. 1780. Mary Terry, aged 60. Had been subject to asthma for several years; after a severe fit of it her legs began to swell, and the quantity of urine to diminish.   In six weeks she was much troubled with the swellings in her thighs and abdomen, which decreased very little when she lay down: she made not quite a pint of water in the twenty-four hours. I ordered her to take two spoonfuls of the infusion of Foxglove every three hours.   By the time she had taken eight doses her urine had increased to the quantity of two quarts in the day and night, but as she complained of nausea, and had once vomited, I ordered the use of the medicine to be suspended for two days.   The nausea being then removed, she again had recourse to it, but at intervals of six hours. The urine continued to discharge freely, and in three weeks she was perfectly cured of her swellings.

C A S E   II.

December, 1782. A poor woman, who had been afflicted with an ague during the whole of her pregnancy, and for two months with dropsical swellings of the feet, legs, thighs, abdomen, and labia pudenda; was at the expiration of the seventh month
taken

taken in labour. On the day after her delivery
the ague returned, with fo much violence as to en-
danger her life. As foon as the fit left her, I began to give her the red bark in fubftance, which
had the defired effect of preventing another pa-
roxyfm. She continued to recover her health for a
fortnight, but did not find any diminution in the
fwellings; her legs were now fo large as to oblige
her to keep conftantly on the bed, and fhe made
very little water. I ordered her the infufion of
Foxglove three times a day, which, on the third
day, produced a very copious difcharge of urine,
without any ficknefs; fhe continued the ufe of it
for ten days, and was then able to walk. Having
loft all her fwellings, and no complaint remaining
but weaknefs, the bark and fteel compleated the
cure.

Extract of a Letter from Doctor Jones, Phyfician, in Lichfield.

ANXIOUS to procure authentic accounts from
the patients, to whom I gave the Foxglove, I have
unavoidably been delayed in anfwering your laft
favour. However, I hope the delay will be made
up by the efficacy of the plant being confirmed by
the enquiry. Long cafes are tedious, and feldom
read, and as feldom is it neceffary to defcribe every fymptom; for every cafe would be a hiftory of
dropfy. I fhall therefore content myfelf with fpeci-
fying

fying the nature of the difeafe, and when the dropfy is attended with any other affection fhall notice
it.

Two years have fcarcely elapfed fince I firft employed the Digitalis; and the fuccefs I have had
has induced me to ufe it largely and frequently.

C A S E   I.

Ann Willott, 50 years of age, became a patient
of the Difpenfary on the 11th of April 1783. She
then complained of an enlargement of the abdomen, difficulty of breathing, particularly when lying, and coftivenefs. She paffed fmall quantities
of high-coloured urine; and had an evident fluctuation in the belly.   Her legs were œdematous.
Chryftals of tartar, fquills, &c. had no effect. The
13th of *June* fhe took two fpoonfuls of a decoction
of Foxglove, containing three drams of the dry
leaves, in eight ounces, three times a day. Her
urine foon increafed, and in a few days fhe paffed
it freely, which continued, and her breath returned.

C A S E   II.

Mr. ———, 45 years of age, had been long
fubject to dropfical fwellings of the legs, and made
little water. Two fpoonfuls of the fame decoction
twice a day, foon relieved him.

CASE

C A S E   III.

Mrs. ——, aged 70 years. A lady frequently afflicted with the gout, and an afthmatical cough. After a long continuance of the latter, fhe had a great diminution of urine, and confiderable difficulty of breathing, particularly on motion, or when lying. Her body was much bound. There was, however, no apparent fwelling. She took three fpoonfuls of an aperient decoction of forty-five grains in fix ounces and a half, every other morning. The urine was plentiful thofe days, and her breathing much relieved. In two or three weeks after the ufe of it fhe was perfectly reftored. The purgative medicine neither increafed the urine, nor relieved the breathing, till the Foxglove was added.

This fpring fhe long laboured with the gout in her ftomach, which terminated in a fit in her hand. During the whole of this tedious illnefs, of nearly three months, fhe paffed little urine, and her breathing was again fhort.

She took the fame preparation of Foxglove without any diuretic effect, and afterwards two and three grains of the powder twice a day with as little. The dulcified fpirits of vitriol, however, quickly promoted the urinary fecretion.

CASE

C A S E   IV.

Mr. C——, 46 years of age, had dropfical fwellings of the legs, and paffed little urine. He took the decoction with three drams, and was foon relieved.

C A S E   V.

Lady ———, took three grains of the dried leaves twice a day, for fwelled legs, and fcantinefs of urine, without effect.

C A S E   VI.

Mrs. Slater, aged 36 years. For dropfy of the belly and legs, and fcantinefs of urine, of feveral weeks ftanding, took three grains of the powder twice a day, and was quite reftored in ten days. She took many medicines without effect.

C A S E   VII.

Mrs. P———, in her 70th year, took three grains of the powder twice a day, for fcantinefs of urine, and fwelled legs, without effect.

C A S E   VIII.

Ann Winterleg, in her 26th year, had dropfical fwellings of the legs, and paffed little urine: fhe was relieved by two drams, in an eight ounce decoction.

C A S E

C A S E   IX.

William Brown, aged 76. In the laſt ſtage of
dropſy of the belly and legs, found a conſiderable
increaſe of his urine by a decoction of Foxglove,
but it was not permanent.

C A S E   X.

Mr.————, — years of age, and of very groſs
habit of body, became highly dropſical, and took
various medicines, without effect. One ounce of
the decoction, with three drams of the dry leaves
in eight ounces, twice or three times a day, increaſed
his urine prodigiouſly. He was evidently better,
but a little attendant nauſea overcame his reſolution,
and in the courſe of ſome weeks afterwards he fell
a victim to his obſtinacy.

C A S E   XI.

Mrs. Smith, about 50 years of age, after a tedious illneſs of many weeks, had a jaundice, and became dropſical in the legs. Two ſpoonfuls of the
decoction, with three drams twice a day, increaſed
her urine, and abated the ſwelling.

C A S E   XII.

Widow Chatterton, about 60 years of age. Took
the decoction in the ſame way for dropſy of the legs,
with little effect.

CASE

C A S E   XIII.

———— Genders, about thirty-four years of age, was delivered of three children, and became dropfical of the abdomen. She paffed little or no urine, had conftant thirft, and no appetite. She took two fpoonfuls of an eight ounce decoction, with three drams twice a day. By the time fhe had finifhed the bottle, (which muft have been on the fourth day,) fhe had evacuated all her water, and could go about. Her appetite increafed with every dofe, and fhe recovered without farther help.

C A S E   XIV.

Mifs M—— M——, in her 20th year. Had been infirm from her cradle, and, after various fufferings, had an aftonifhing œdematous fwelling of one leg and thigh, of many weeks ftanding. She paffed little or no urine, and had all her other complaints. She took 2 fpoonfuls of an eight oz. decoction of two drams, twice a day. Her urine immediately increafed; and, on the third day, the fwelling had entirely fubfided.

C A S E   XV.

Mr. P——, 65 years of age, and of a full habit of body. Had lived freely in his youth, and for many years led rather an inactive life. His health was much impaired feveral months, and he had a confiderable diftention, and evident fluctuation in

I

the

the abdomen, and a very great œdema of the legs and thighs. His breathing was very fhort, and rather laborious, appetite bad, and thirft confiderable. His belly was bound, and he paffed very fmall quantities of high-coloured urine, that depofited a reddifh matter. He had taken medicines fome time, and, I believe, the Digitalis; and had been better.

A blifter was applied to the upper and infide of each thigh; he took two fpoonfuls of the decoction, with three drams of the dry leaves, two or three times a day; and fome opening phyfic occafionally.

He lived at a confiderable diftance, and I did not vifit him a fecond time; but I was well informed, about ten days or a fortnight afterwards, that his urine increafed amazingly upon taking the decoction, and that the water was entirely evacuated.

C A S E   XVI.

Mrs. G——, aged 50 years. After being long ailing, had a large collection of water in the abdomen and lower extremities. Her urine was high-coloured, in fmall quantities, and had a reddifh fediment. She took the decoction of Digitalis, fquills, &c. without any effect. The chryftals of tartar, however, cured her fpeedily.

CASE

C A S E   XVII.

Mr. ———, about 50 years of age, complained of great tenfion and pain acrofs the abdomen, and of lofs of appetite; his urine, he thought, was lefs than ufual, but the difference was fo trifling he could fpeak with no certainty: his belly feemed to fluctuate. Among other things he tried the Fox-glove leaves dried, twice a day; and, although it appeared to afford him relief, yet the effect was not permanent.

C A S E   XVIII.

Mr. W———, aged between 60 and 70 years; and rather corpulent: was confiderably dropfical, both of the belly and legs, and his urine in fmall quantities. Three grains of the dry leaves, twice a day, evacuated the water in lefs than a fortnight.

C A S E   XIX.

Sarah Taylor, 40 years of age, was admitted into the Difpenfary for dropfy of the abdomen and legs; and was relieved by the Decoctum digitalia-num.

C A S E   XX.

Lydia Smith, aged 60. Difpenfary. Laboured many years under an afthma, and became dropfi-cal. She took the decoction without effect.

CASE

C A S E   XXI.

John Leadbeater, aged 15 years. Had a quoti-
dian intermittent, which was removed by the hu-
mane affiftance of an amiable young lady. His
intermittent was foon attended by a very confiderable afcites; for which he became a patient of the
Difpenfary. He took a decoction of Foxglove night
and morning. His urine increafed immediately,
and he loft all his complaints in four days.

C A S E   XXII.

William Millar, aged 50 years. Admitted into
the Difpenfary for a tertian ague, and general dropfy. The dropfy continuing after the ague was removed, and his urine being ftill paffed in fmall
quantities; he took the powdered leaves, and recovered his health in five days.

C A S E   XXIII.

Ann Wakelin, 10 years of age. Had for feveral weeks a dropfy of the belly after an ague. She
took a decoction of Foxglove, which removed all
complaint by the fourth day.

C A S E   XXIV.

Ann Meachime; a Difpenfary patient. Had an
afcites and fcantinefs of urine. She took the powder

der of Foxglove, and evacuated all her water in three days.

It may not be improper to obferve, 1ft. That various diuretics had long been given in many of thefe cafes before I was confulted. And, 2dly. That the exhibition of the Foxglove was but feldom attended with ficknefs.

R E M A R K S.

Thefe Cafes, thus liberally communicated by my friend, Dr. Jones, are more acceptable, as they feem to contain a faithful abftract from his notes, both of the unfuccefsful as well as the fuccefsful Cafes.

The following Tabular View of them will give us fome Idea of the efficacy of the Medicine.

Anafarca - - - - , 7 Cafes - Cured - - 3
                                 Relieved - 1
                                 Failed - - 3
Afcites - - - - 5 Cafes - Cured - - 4
                                 Relieved - 1
Œdematous leg - - 1 Cafe - Cured - - 1
Afcites and anafarca - 7 Cafes - Cured - - 4
                                 Relieved - 2
                                 Failed - - 1
Afthma and dropfy - 1 Cafe - Failed - - 1
Hydrothorax and gout 1 Cafe - Cured - - 1
- - - - -, afcites }
  and anafarca - - } 2 Cafes - Cured - - 2

I 3   A CASE

A CASE of Anafarca communicated by Mr.
Jones, Surgeon, in Birmingham.

Dear Sir,

HAVING lately experienced
the diuretic powers of the Foxglove, in a cafe of
anafarca; I do myfelf the pleafure of communicating a fhort hiftory of the treatment to you.

I am, &c.

Birmingham,   W. JONES.
May 17th, 1785.

My patient, Mrs. C———, who is in her 51ft
year, had the following fymptoms, viz. alternate
fwelling of the legs and abdomen, a little cough,
fhortnefs of breath in a morning, thirft, weak pulfe,
and her urine, which was fo fmall in quantity as
feldom to amount to half a pint in twenty-four
hours, depofited a clay-coloured fediment.

April 16th, 1785, I directed the following form:

R. Fol. Digitalis ficcat. ʒii.
    Aq. fontanæ bullient. ʒviii. f. infuf. et cola.
    Sumat cochl. larga iii. o. n. et mane.

On the 17th fhe had taken twice of the infufion,
and though by miftake only two tea fpoonfuls for a
dofe,

dofe, yet the quantity of urine was increafed to about a pint in the twenty-four hours. She was then directed to take two table fpoonfuls night and morning. And,

On the 18th, a degree of naufea was produced. A pint and half of urine was made in the laft twenty-four hours. During the time above fpecified fhe had two or three ftools every day. The infufion was now omitted.

On the 19th the fwelling of the legs was removed. A degree of naufea took place in the morning, and increafed fo much during the day, that fhe vomitted up all her food and medicine. As fhe was very low, and complained of want of appetite, a cordial julep was directed to be taken occafionally, as well as red port and water, mint tea, &c. She informed me that whatever fhe took generally ftaid about an hour before it came up again, and that the mint tea ftaid longeft on the ftomach. The vomiting decreafed gradually, and ceafed on the 22d. The difcharge of urine remained confiderable during the three following days, but its quantity was not meafured.

22d. A dofe of neutral faline julep was directed to be taken every fourth hour.

On the 23d fhe complained of thirft, and thought the difcharge of urine not fo copious as on the preceding days, therefore the faline julep was continued ed every fourth hour, with the addition of thirty drops of the following medicine:

R. Aceti fcillitic. ʒvi.
   Tinct. aromat. ʒii.
   Tinct. thebaic. gutt. xx. m.

The bowels have been kept open from the 19th, by the occafional ufe of emollient injections.

On the 24th the legs were much fwelled again; fhe complained of languor and a degree of naufea. The difcharge of urine increafed a little fince the 23d. Her pulfe was low and her tongue white. The urine, which had been rendered clear by the infufion of Foxglove, now depofited a whitifh fedi-ment.

On the 25th her appetite began to return, the fwelling of the legs diminifhed, and fhe thought herfelf much relieved. The urine was confiderable in quantity, and clear.

On the 26th fhe was thirfty and languid. The fwelling was removed; the quantity of urine dif-charged in the laft twenty-four hours was about a pint. She continued to mend from this time, and is now in good health,

A giddinefs of the head, more or lefs remarkable at times, was obferved to follow the ufe of the Fox-glove, and it lafted nine or ten days.

This

This is the fecond time that I have relieved this patient by the infufion of Foxglove.   I ufed the fame proportion of the frefh leaves the firft time as I did of the dried ones the laft.   The violent vomiting which followed the ufe of the infufion made with the dried leaves, did not take place with the frefh, though fhe took near a pint made with the fame proportion of the herb frefh gathered.

R E M A R K S.

T H E above is a very inftructive cafe, as it teaches us how fmall a quantity of the infufion was neceffary to effect every defirable purpofe.   At firft fight it may appear from the concluding paragraph, that the green leaves ought to be preferred to the dried ones, as being fo much milder in their operation; but let it be noticed, that the fame quantity of infufion was prepared from the fame weight of the green as of the dried leaves, and confequently, as will appear hereafter, the infufion with the dried leaves was five times the ftrength of that before prepared from the green ones.   We need not wonder, therefore, that the effects of the former were fo difagreeable, when the dofe was five times greater than it ought to have been.   But what makes this matter ftill more obvious, is the miftake mentioned at firft, of two tea fpoonfuls only being given for a dofe.   Now a tea fpoonful, containing about a fourth or a fifth part of the contents of a table fpoon, the dofe then given, was very nearly the fame as that which had before been taken of the

infufion

infusion of the green leaves, and it produced pre-
cisely the same effects for it increased the urinary
discharge, without exciting the violent vomiting.

Letter from Doctor Johnstone, Physician, in Birmingham.

Dear Sir,

THE following cases are selected
from many others in which I have given the D gitalis purpurea; and from repeated experience of its
efficacy after other diuretics have failed, I can re-
commend it as an effectual, and when properly
managed, a safe medicine.

I am, &c.

Birmingham, May 26,   E. JOHNSTONE.
    1785.

March 8th, 1783, I was called to attend Mr.
G——, a gentleman of a robust habit, who had led
a regular and temperate life, Æt. 68.   He was
affected with great difficulty of respiration, and cough
particularly troublesome on attempting to lie down,
œdematous swellings of the legs and thighs, abdomen tense and sore on being pressed, pain striking
from the pit of the stomach to the back and shoul-
ders; almost constant nausea, especially after taking
food, which he frequently threw up; water thick
and high-coloured, passed with difficulty and in
                                            small

fmall quantity; body coftive; pulfe natural; face
much emaciated, eyes yellow and depreffed. He
had been fubject to cough and difficulty of breathing
in the winter for feveral years; and about four
years before this time, after being expofed to cold,
was fuddenly deprived of his fpeech and the ufe of
the right fide, which he recovered as the warm wea-
ther came on ; but fince that time had been remark-
ably coftive, and was in every refpect much debilitated. He firft perceived his legs fwell about a year
ago ; by the ufe of medicines and exercife, the
fwellings fubfided during the fummer, but returned
on the approach of winter, and gradually increafed
to the ftate in which I found them, notwithftanding he had ufed different preparations of fquills and
a great variety of other diuretic medicines. I
ordered the following mixture.

> R. Foliorum Digitalis purpur. recent. ʒiii. deco-
> que ex aq. fontan. ℥xii ad ℥vi colaturæ adde
> Tinctur. aromatic.
> Syr. zinzib. aa ℥i. m. capt. cochl. duo larga fe-
> cunda quaque hora ad quartam vicem nifi
> prius naufea fupervenerit.

March 9th. He took four dofes of the mixture
without being in the leaft fick, and made, during
the night upwards of two quarts of natural coloured
water.

10th.

10th. Took the remainder of the mixture yefterday afternoon and evening, and was fick for a fhort
time, but made nearly the fame quantity of-water
as before, the fwellings are confiderably diminifhed,
his appetite increafed, but he is ftill coftive.

> R. Argent. viv. balfam peruv. aa ʒfs tere ad ex-
> tinctionem merc. et adde gum. ammon.
> Ɔiii aloes focotorin. 3fs rad. fcil. recent. Ɔfs
> fyr. fimpl. q. f. f. mafs. in pil. xxxii divid.
> cap. iii. bis in die.

14th. Continues to make water freely.   The
fwellings of his legs have gradually decreafed ; forenefs and tenfion of the abdomen confiderably lefs.

> Omittant. pil. cap. miftur. c. decoct. Digitalis &c,
> 3tia quaque hora ad 3tiam vicem.

15th. Made a pint and a half of water laft night,
without being in the leaft fick, and is in every
refpect confiderably better.   Repet. Pillul. ut
antea.

21ft. Makes water as ufual when in health, and
the fwellings are entirely gone.

> R. Infus. amar. ʒv. tinctur. Rhei fpirit. Ʒii. fpi-
> rit vitriol. dulc. ʒii. fyr. zinzib. ʒvi. m. cap.
> cochl. iii. larg. ter in die.

He foon gained fufficient ftrength to enable him
to go a journey, and returned home in much better
health

health than he had been from the time he was affected with the paralytic ftroke, and excepting fome return of his afthmatic complaint in the winter, hath continued fo ever fince.

C A S E   II.

R—— Howgate, a man much addicted to intemperance, particularly in the ufe of fpirituous liquors, Æt. 60, was admitted into the Hofpital near Birmingham, *May* 17, 1783. He complained of difficulty of breathing, attended with cough, particularly troublefome on lying down; drowfinefs and frequent dozing, from which he was roufed by ftartings, accompanied with great anxiety and oppreffion about the breaft ; œdematous fwellings of the legs; conftant defire to make water, which he paffed with difficulty, and only by drops ; pulfe weak and irregular; body rather coftive ; face much emaciated; no appetite for food.—Cap. pil. fcil. iii. ter in die.*

May 20th. The pills have had no effect.—Cap. miftur. c. † Decoct. Digital. &c. cochl.ii. larg. 3tia quaque hora, ad 3tiam vicem.

May 21ft. Made near two quarts of water in the night, without being in the leaft fick. He continued
the

---

* R. Rad. fcil. recent. fapon. caftiliens. pulv. Rhei opt. aa. Ði. ol. junip. gutt. xvi. fyr. balf. q. s. f. mafs. in pil. xxiv. divid.

† Prepared in the fame manner as in the former cafe.

the ufe of the mixture three times in the day till the 30th, and made about three pints of water daily, by which means the fwellings were entirely taken away ; and his other complaints fo much relieved, that on the 6th of June he was difmiffed free from complaint, except a flight cough. But returning to his old courfe of life, he hath had frequent attacks of his diforder, which have been always removed by ufing the Digitalis.

Extract of a letter from Mr. Lyon, Surgeon, at Tamworth.

—Mr. Moggs was about 54 years of age, his dif- eafe a dropfy of the abdomen, attended with anafarcous fwellings of the limbs, &c. brought on by exceffive drinking. I believe the firft fymptoms of the difeafe appeared the beginning of November, 1776 ; the medicines he took before you faw him, were fquills in different forms, fal diureticus and calomel, but without any good effect ; he begun the Digitalis on the 10th of July 1777 ; a few dofes of it caufed a giddinefs in the head, and almoft de- prived him of fight, with very great naufea, but very little vomiting, after which a confiderable flow of urine enfued, and in a very fhort time, a very little water remained either in the cavity of the abdomen, or the membrana adipofa, but he remained exceffive weak, with a fluttering pulfe at the rate of 150 or frequently 160 in a minute ; he kept pretty free from water for upwards of twelve months ; it then collect-

collected, and neither the Digitalis nor any other medicine would carry it off. I tapped him the 2d of Auguſt 1779 in the uſual place, and took ſome gallons of water from him, but he very ſoon filled again, and as he had a very large rupture, a conſiderable quantity of the water lodged in the ſcrotum, and could not be got away by tapping in the uſual place. I therefore (on the 28th of the ſame month) made an inciſion into the lower part of the ſcrotum, and drained off all the water that way, but he was ſo very much reduced, that he died the 8th or 9th of *September* following, which was about two years and two months after he firſt begun the Digitalis.

I have had ſeveral dropſical patients relieved, and ſome perfectly recovered by the Digitalis, ſince you attended Mr. Moggs, but as I did not take any notes or make any memorandums of them, cannot give you any of them.

Communications from Dr. Stokes, Phyſician, in Stourbridge.

Dear Sir,

        I ACCEPT with pleaſure your invitation to communicate what I know reſpecting the properties of *Digitalis*; and if an account of what others had diſcovered before you,* with a detail

---

* See this account in the Introduction.

tail of my own experience, fhall be allowed the merit of at leaft a well meant acknowledgment, for the early communication you were fo kind to make me, of the valuable properties you had found in it; I fhall confider my time as well employed. A knowledge of what has been already done is the beft ground work of future experiment; on which account I have been the more full on this fubject, in hopes that given with the cautions which you mean to lay down in the cure of dropfies, it may prove alike ufeful in that of other difeafes, one of which ftands foremoft among the *opprobria* of medicine.

C A S E I.

Mrs. M——. Orthopnea, pain, and exceffive oppreffion at the bottom of the fternum. Pulfe irregular, with frequent intermiffions. Appetite very much impaired. Legs anafarcous.

Empl. veficator. pectori dolent.
Infuf. Digital. e ʒiii. ad. aq. &c. ʒviii. cochl. j. o.
h. donec naufea excitetur vel diurefis fatis copiofa proveniat.

I ordered it of the above ftrength, and to be repeated often, on account of the great emergency of the cafe, but the naufea excited by the firft dofe prevented its being given at fuch fhort intervals. A 3d dofe I found had been given, which was followed by vomitings. All her complaints gradually abated, but

but in about a fortnight recurred, notwithſtanding the uſe of infuſ. amar. &c.

Dec. 2. *Infus. Digit. e. ʒiſs ad aq. &c. ℥viii. cochl. ii. horis &c. u. a.*

Complaints gradually abated, ſwellings of the legs nearly gone down.

About a month afterwards you was deſired to viſit this patient.*

---

* For reaſons aſſigned at p. 100, I did not intend to introduce any caſe, occuring under my own inſpection, in the courſe of the preſent year ; but it may be ſatisfactory to continue the hiſtory of this diſeaſe, as Dr. Stokes's narrative would otherwiſe be incomplete.

1785.

C   A   S   E.

Jan. 5th. Mrs. M——, Æt. 48. Hydrothorax and anaſarcous legs, of eight months duration. She had taken jallap, ſquill, ſalt of tartar, and various other medicines. I found her in a very reduced ſtate, and therefore directed only a grain and half of the Pulv. Digital. to be given night and morning. This in a few days encreaſed the ſecretion of urine, removed her difficulty of breathing, and reduced the ſwelling of her legs, without any diſturbance to her ſyſtem.

Three months afterwards, a ſevere attack of gout in her legs and arms, removing to her head, ſhe died.

Dr. Stokes had an opportunity of examining the dead body, and I had the ſatisfaction to learn from him, that there did not appear to have been any return of the dropſy.

K   On

On the examination of the body I noticed, among others, the following appearances.

About ¾ oz. of bloody water flowed out, on elevating the upper half of the fcull, and a fmall quantity alfo was found at the bafe.

BRAIN. Blood-veffels turgid with blood, and many of thofe of confiderable fize diftended with air.

A very flight watery effufion between the *Pia Mater* and *Tunica arachnoidea*. About ¾ oz. of watery fluid in the *lateral ventricles*.

THORAX. In the left cavity about 4 oz. of bloody ferum; in the right but little. Lungs, the hinder parts loaded with blood. Adhefions of each lobe to the pleura. *Pericardium* containing but a very fmall quantity of fluid. *Heart* containing no coagula of blood. *Valves of the Aorta* of a cartilaginous texture, as if beginning to offify.

Abdominal Vifcera natural, and a profufion of *Fat* under the integuments of the abdomen and thorax, in the former to the thicknefs of an inch and upwards, and in very confiderable quantity on the mefentery, omentum, kidneys, &c.

OBS. The intermitting pulfe fhould feem to have been owing to effufions of water in fome of the cavities of the breaft, as it difappeared on the removal of the waters.

CASE

CASE II.

Mrs. C—— of K————, Æt. 80. Orthopnœa, with fenfe of oppreffion about the prœcordia. Unable to lie down in bed for fome nights paft. Anafarca of the lower extremities. Urine very fcanty. Complaints of fix weeks ftanding. Had taken *fal. diuret. c. ol. junip.*—*Calom.c. jalap, et gambog.*—*Et ol. junip. c. ol. Terebinth.* without effect.

Feb. 7. *Infuf. Digital. e. ʒiii. ad aq. &c. ʒviii. cochl. ii. 4tis horis.* Ordered to drink largely of *infus. baccar. junip.* The third dofe produced great naufea which continued ten hours, during which time the urine made was about a quart. The next day her apothecary directed her to begin again with it. The fecond dofe produced vomiting. During the next twenty hours fhe made two quarts of water, about four times as much as fhe drank.

From this time fhe took no more of the *infus. Digital.* but continued the *inf. bacc. junip.* until about *March* 2d, when all the fwellings were gone down, her refpiration perfectly free, and fhe herfelf quite reftored to her former ftate of health. On the 29th fhe had an attack of jaundice which was fome time after removed; fince which fhe has enjoyed a good ftate of health, excepting that for fome little time paft her ancles have been flightly œdematous, which will I truft foon yield to ftrengthening medicines.

CASE

C A S E III.

Mrs. M—— G——, Æt. 64. Has had fore legs for thefe thirty-four years paft. Orthopnœa. Senfe of oppreffion at the præcordia. Pulfe intermitting. Legs anafarcous. Urine fcanty, high-coloured.

Infus. Digital. c. ʒifs ad aq. bull. ʒviii. cochl. ii. 4tis horis.

Took fix dofes, when naufea was excited. Urine a quart during the courfe of the night. The flow of urine continued, and complaints relieved. Sal. Mart. c. extr. gent. and afterwards with the addition of extr. cort. for which laft ingredient fhe had a predilection, confirmed the cure.

On the fame day the next year I was called in to her for a fimilar train of fymptoms, excepting that the pulfe was but juft perceptibly irregular.

Infus. Digital. u. a. præfcript.

The directions on the phial not being attended to, *two dofes of it were given after a naufea had been excited*, which, with occafional vomitings, became exceedingly oppreffive. A faline draught, given in Dr. Hulme's method, a draught *fal. c.. c. gr. xii. c. conf. card. gr. x.* produced no immediate effect, but the naufea gradually abating, inf. bacc. junip. was ordered; but this appeared to augment it, and

and a great propenfity to fleep coming on, I directed *fal. c. c. conf. card. aa gr. viii. 4tis horis*, which
removed the unpleafant fymptoms and *myrrh. c. fal.
mart.* completed the cure.   During the ufe of the
above medicines, the urine was augmented, and the
pulmonary complaints removed, even before the naufea left her; and the fores of her legs which were much
inflamed before fhe began with the infuf. Digital.
in a day's time affumed a much healthier appearance, and on her other complaints going off, they
fhewed a greater tendency to heal than fhe had
ever obferved in them for twenty years before. This
inftance is a very pleafing confirmation of the experience of Hulfe and Dr. Baylies, and of the advantage to be derived from a medicine, which,
while it helps to heal the ulcers, removes that from
the conftitution which often renders the healing of
them improper.

In one cafe in which I ordered it, the infufion,
inftead of digefting three hours as I had directed,
was fuffered to ftand upon the leaves all night. The
confequence was that the firft dofe produced confiderable naufea.

The two following cafes, with which I have been
favoured by a phyfician very juftly eminent, con-
vince me of the neceffity there is that every one who
difcovers a new medicine, or new virtues in an old
one, fhould, in announcing fuch difcoveries, publifh
to the world the exact manner in which he exhi-

bits

bits fuch medicines, with all the precautions neceffary to obtain the promifed fuccefs.

In thefe (fays my correfpondent) " the infufion
" was given in fmall dofes, repeated every hour or
" two, till a naufea was raifed, when it was omit-
" ted for a day or perhaps two, and then repeated
" in the fame manner."

" An Ascites emptied by it, but filled again
" very fpeedily, though *its ufe was never difconti-*
" *nued*, and who afterwards found no falutary ef-
" fects from it. Ended fatally."

" In an Anasarca it fometimes increafed the
" quantity of urine, and abated the fwelling, but
" which as often returned in as great a degree as
" before, though *the medicine was ftill given*, and al-
" ways increafed in quantity fo as to excite naufea.
" Ended fatally."

" I have tried it in many other cafes, but found
" very little difference in the fuccefs attending it."

May we not be allowed to conjecture that the inefficacy of *its continued ufe* is owing to its narcotic pro-
perty gradually diminifhing the irritability of the
mufcular fibres of the abforbents, or poffibly of the
whole vafcular fyftem, and thus adding to that
weakened action which feems to be the caufe of the
generality of dropfies, which leads us to caution
the medical experimenter againft trying it, at leaft
againft

againſt its continued uſe, even in ſmall doſes, in other diſeaſes of diminiſhed energy, as continued fever, palſy, &c.

I remain with the greateſt truth,

Your obliged and affectionate friend,

Stourbridge,   JONATHAN STOKES.
May 17, 1785.

T H E three following Hoſpital Caſes, which Dr. STOKES had an opportunity of obſerving, are related as inſtances of bad practice, and tend to demonſtrate how necefſary it is when one phyſician adopts the medicine of another, that he ſhould alſo at firſt rigidly adopt his method.

C A S E   I.

Eſther K——, Æt. 33. General anaſarca, aſcites, and dyſpnœa, of ſeven months duration.

Decoct. e Digit. ʒiv. c. aq. ℔i. coquend. ad ℔ſs. cap. ʒi. 2dis. horis. 1ſt DAY. 4th doſe made her ſick. 2d DAY. The firſt doſe ſhe took to-day produced vomiting.

3d DAY.

3d Day. *Minuatur dofis ad ℥ſs.* This ſtayed upon her ſtomach, but produced an almoſt conſtant ſickneſs. Stools more frequent, water ſcarce ſenſibly increaſed; and her ſwellings not at all reduced.

4th Day. *Cap. Calomel. gambog. ſcill. &c.*

Obs. Sufficient time was not allowed to obſerve its effects, neither was the patient enjoined the free uſe of diluents. The diſeaſe terminated fatally.

CASE II.

William T——, Æt. 42. Aſcites, with cough and dyſpnœa. Abdomen very much diſtended. The reſt of his body highly emaciated. Urine thick, high coloured, and in very ſmall quantity.

Decoct Digit. (u. in Eſther K——,) 4tis horis.

1ſt Day of taking it. The 4th doſe produced ſickneſs.

2d. Vomiting after the ſecond doſe.

10th. Urine increaſed to ℔vi.

11th. Flow of urine continues. Abdomen quite flaccid.

12th. Ab-

12th. Abdomen not diminifhed.

15th. A fmart purging came on, and the flow of urine diminifhed.

23d. Belly much bound. Took a cathart. powder, which was followed by a diminution of the abdomen.

29th. To take a cathart. powder every 4th morning, continuing the decoct. Digit.

32d. Urine exceedingly fcanty.

35th. *Vin. fcill.* ℥*fs. o. m. &c.* This produced diuretic effects.

44th. Tapped. Terminated fatally.

OBS. Here the medicine was *continued till it ceafed to produce diuretic effects;* and thefe effects were not aided by any ftrengthening remedies.

C A S E   III.

George R——, Æt. 52. Afcites, general anafarca, and dyfpnœa. His legs fo greatly diftended that it was with great difficulty he could draw the one after the other.

Infuf.

Infuf. Digital. ʒiiifs. ad. aq. ℔fs. cap. ℥i. altern. horis donec naufeam excitaverit. Rep. 3tiis diebus. tempore intermedio cap. fol. guaic. ℥i. ter in die ex inf. finap.

1ſt DAY of taking it. Became ſickiſh towards night.

2d DAY. Made a great quantity of water during the night, and ſpat up a great deal of watery phlegm. The firſt doſe he took in the morning has produced a ſickneſs which has continued all day, but he has never vomited.

3d. DAY. The change in his appearance ſo great as to make it difficult to conceive him to be the ſame perſon. Inſtead of a large corpulent man, he appeared tall, thin, and rather aged. Breathe. freely, and can walk up and down ſtairs without inconvenience.

4th DAY. *Decoɛ̃. bacc. junip. and cyder for common drink.*

6th DAY. A ſecond courſe of his medicine produced a flow of urine almoſt as plentiful as the former, though he drank little or nothing at the time. In a day or two after he walked to ſome diſtance.

12th DAY. *Pot. purgans illico.*

14th DAY. *Pot. purg. c. jalap. ʒfs. 4tis diebus. Infuf. Dig. 3tiis diebus.*

17th DAY.

17th Day. *R. Gamb. gr. iii. calom. gr. ii. camph. gr. i. ſyr. ſimpl. fiat pil. o. n. ſum. Infuſ. Digit. 3tiis diebus.*

21ſt Day. Made an out-patient. The ſuper-abundant flow of urine continued for the firſt three days after his laſt courſe; but ſince, the flow of ſaliva has been nearly equal to that of urine.

The ſmalls of his legs not quite reduced, and are fuller at night. He has ſhrunk round the middle from four feet two inches to three feet ſix inches; and in the calves of his legs, from ſeventeen inches to thirteen and a half.*

Obs. The waters were here very ſuccefsfully evacuated, but as you remarked to me, on communicating the caſe to you at the time, tonic medicines ſhould have been given, to ſecond the ground that had been gained, inſtead of weakening the patient by draſtic purgatives.

---

* In the three laſt recited caſes, the medicine was directed in doſes quite too ſtrong, and repeated too frequently. If Eſther K—— could have ſurvived the extreme ſickneſs, the diuretic effects would probably have taken place, and, from her time of life, I ſhould have expected a recovery. Wm. T —— ſeems to have been a bad caſe, and I think would not have been cured under any management. G. R—— certainly poſſeſſed a good conſtitution, or he muſt have ſhared the fate of the other two.

A CASE

A C A S E from Mr. Shaw, Surgeon, at Stourbridge. — Communicated by Doctor Stokes.

Matth. D——, Æt. 71. Tall and thin. Difeafe a general anafarca, with great difficulty of breathing. The lac ammoniac. fomewhat relieved his breath ; but the fwellings increafed, and his urine was not augmented. I confidered it as a loft cafe, but having feen the good effects of the Digitalis, as ordered by Dr. Stokes in the cafe of Mrs. G——, I gave him one fpoonful of an infufion of ʒii. to half a pint, twice a day. His breath became much eafier, his urine increafed confiderably and the fwellings gradually difappeared; fince which his health has been pretty good, except that about three weeks ago, he had a flight dyfpnœa, with pain in his ftomach, which were foon removed by a repetition of the fame medicine.

Mr. Shaw likewife informs me, that he has removed pains in the ftomach and bowels, by giving a fpoonful of the infufion, ʒifs. to ʒviii. morning and night.

A Letter

A Letter from Mr. V A U X, Surgeon, in Birmingham.

Dear SIR,

 I SEND you the two following cafes, wherein the Digitalis had very powerful and fenfible effects, in the cure of the different patients.

CASE I.

 Mrs. O—— of L—— ftreet, in this town, aged 28, naturally of a thin, fpare habit, and her family inclinable to phthifis, fent for me on the 11th of June, 1779, at which time fhe complained of great pain in her fide, a conftant cough, expectorated much, which funk in water; had colliquative fweats and frequent purging ftools; the lower extremities and belly full of water, and from the great difficulty fhe had in breathing, I concluded there was water in the cheft alfo.   The quantity of water made at a time for three weeks before I faw her, never amounted to more than a tea-cup full, frequently not fo much. Finding her in fo alarming a fituation, I gave it as my opinion fhe could receive no benefit from medicine, and requefted her not to take any; but fhe being very defirous of my ordering her fomething, I complied, and fent her a box of gum pills with fquills, and a mixture with falt of tartar: thefe medicines fhe took until the fixteenth, without any good effects : the water in her legs now began to ex-

fude

fude through the fkin, and a fmall blifter on one of
her legs broke. Believing fhe could not exift much
longer, unlefs an evacuation of the water could be
procured; after fully iuforming her of her fituation,
and the uncertainty of her furviving the ufe of the
medicine, I ventured to propofe her taking the Digitalis, which fhe chearfully agreed to. I accordingly fent her a pint mixture, made as under, of
the frefh leaves of the Digitalis. Three drams infufed in one pint of boiling water, when cold ftrained off, without preffing the leaves, and two ounces
of the ftrong juniper water added to it : of this
mixture fhe was ordered four table fpoonfulls every
third hour, till it either made her fick, purged her,
or had a fenfible effect on the kidneys. This mixture was fent on the feventeenth, and fhe began
taking it at noon on the eighteenth. At one o'clock
the following morning I was called up, and informed fhe was dying. I immediately attended
her, and was agreeably furprifed to find their fright
arofe from her having fainted, in confequence of the
fudden lofs of twelve quarts of water fhe had made
in about two hours. I immediately applied a roller
round her belly, and, as foon as they could be made, 2
others, which were carried from the toes quite up the
thighs. The relief afforded by thefe was immediate;
but the medicine now began to affect her ftomach fo
much, that fhe kept nothing on it many minutes together. I ordered her to drink freely of beef tea, which
fhe did, but kept it on her ftomach but a very fhort
time. A neutral draught in a ftate of effervefcence was
taken to no good purpofe : She therefore continued
the.

the beef tea, and took no other medicine for five
days, when her fickncfs went off : her cough abated,
but the pain in her fide ftill continuing, I applied a
blifter which had the defired effect : her urine after
the firft day flowed naturally. Her cure was compleated by the gum pills with fteel and the bitter
infufion. It muft be obferved fhe never had any
collection of water afterwards.

It affords me great pleafure to inform you that
fhe is now living, and has fince had four children ;
all of whom, I think I may juftly fay, are indebted
to the Digitalis for their exiftence.

There appears in this cafe a ftriking proof of the
utility of emetics in fome kinds of confumptions, as
it appears to me the dropfy was brought on by the
cough, &c. and I believe thefe were cured by the
continual vomitings, occafioned by the medicine.

C A S E II.

Mr. H——, a publican, aged about 48 years,
fent for me in *March*, 1778. He complained of a
cough, fhortnefs of breathing, which prevented him
from laying down in bed ; his belly, thighs and legs
very much diftended with water ; the quantity of
urine made at a time feldom exceeded a fpoonful.
I requefted him to get fome of the Digitalis, and as
they had no proper weights in the houfe, I told them
to put as much of the frefh leaves as would weigh
down a guinea, into half a pint of boiling water ;

to

to let it ſtand till cold, then to pour off the clear liquor, and add a glaſs of gin to it, and to take three
table ſpoonfuls every third hour, until it had ſome
ſenſible effect upon him.

Before he had taken all the infuſion, the quantity of urine made increaſed, (he therefore left off
taking it), and it continued to do ſo until all the
water was evacuated.   His breathing became much
better, his cough abated, though it never quite left
him ; he being for ſome time before aſthmatic.
By taking ſome tonic pills he continued quite well
until the next ſpring, when he had a return of his
complaint, which was carried off by the ſame means.
Two years after, he had a third attack, and this alſo gave way to the medicine.   Laſt year he died of
a pleuriſy.

I am, &c.

Moor-Street, 8th May,   JER. VAUX.
      1785.

P. S. You muſt well recollect the caſe of Mrs.
F——. —It was " a general dropſy—every time
" ſhe took the medicine its effects were ſimilar, viz.
" The diſcharge of urine came on gradually at firſt,
" increaſed afterwards, and the whole of the water
" both in the belly, legs,&c. was perfectly evacuated.
" Although the effects were only temporary, they
" were exceedingly agreeable to the patient, making
" her time much more comfortable."—— (See Caſe
XLIII.)

A Let-

A Letter from Mr. WAINWRIGHT, Surgeon, in Dudley.

Dear Sir,

        I T gives me great pleafure to find you intend to publifh your obfervations on the Digitalis purpurea.

Several years are now elapfed fince you communicated to me the high opinion you entertained of the diuretic qualities of this noble plant. To enfure fuccefs, due attention was recommended to its *preparation*, its *dofe*, and its *effects* upon the fyftem.

I always gave the infufion of the dried leaves; the dofe the fame as in the prefcriptions returned. If the medicine operated on the ftomach or bowels, it was thought prudent to forbear. When the kidneys began to perform their proper functions, and the urine to be difcharged, a continuance of its farther ufe was unneceffary.

Thefe remarks you made in the cafe of the firft patient for whom you prefcribed the Digitalis in our neighbourhood, and I have found them all neceffary at this prefent period. From the *decided* good effects that followed from its ufe, in thofe cafes where the moft powerful remedies had failed, I was foon convinced it was a moft valuable addition to the materia medica.

L   The

The want of a certain diuretic, has long been one of the defiderata of medicine. The Digitalis is undoubtedly at the head of that clafs, and will feldom, if properly adminiftered, difappoint the expectation. I can fpeak with the more confidence, having, in an extenfive practice, been a happy witnefs to its good qualities.

For feveral years, I have given the infufion in a variety of cafes, where there was a deficiency in the fecretion of the urine, with the greateft fuccefs. In recent obftructions, I do not recollect many failures. In anafarcous difeafes, and in the anafarca, when combined with the afcites ; in fwellings of the limbs, and in difeafes of the cheft, when there was the greateft reafon to believe an accumulation of ferum, the moft beneficial confequences have followed from its ufe.

Had I been earlier acquainted with your intention to publifh an account of the Digitalis, I could have tranfmitted fome cafes, which might have ferved to corroborate thefe affertions : but I am convinced the Digitalis needs not my affiftance to procure a favorable reception. Its own merit will enfure fuccefs, more than a hundred recited cafes.

I could wifh thofe gentlemen who intend to make ufe of this plant, to collect it in a hot dry day, when the petals fall, and the feed-veffels begin to fwell.

The

The leaves kept to the second year are weaker, and their diuretic qualities much diminished. It will therefore be neceffary to gather the plant frefh every feafon.

Thefe cautions are unneceffary to the accurate botanift, who well knows, that a plant in the fpring, though more fucculent and full of juices, is deftitute of thofe qualities which may be expected when that plant has attained its full vigour, and the feed-veffels begin to be manifeft. But for want of attention to thefe particulars, its virtues may be thought exaggerated, or doubtful, if beneficial confequences do not always flow from its ufe. There are difeafes it cannot cure ; and in feveral of thofe patients in this town, who firft took the Digitalis by your orders, there was the moft pofitive proof of the vifcera being unfound. In thefe defperate cafes it often procured a plentiful flow of urine, and palliated a difeafe which mecine could not remove.

At a remote diftance, phyficians are feldom applied to for advice in trifling diforders. Many remedies have been tried without relief, and the difeafe is generally obftinate or confirmed. — It would not be fair to try the merits of the Digitalis in this fcale. It might often fail of promoting the end defired. I flatter myfelf the reputation of this plant will be equal to its merit, and that it will meet with a candid reception.

As

As there is no pleasure equal to relieving the miseries and distresses of our fellow-creatures, I hope you will long enjoy that peculiar felicity.

Permit me to return my thankful acknowledgments, for your free communication of a medicine, by which means, through the blessing of providence, I have been enabled to restore health and happiness to many miserable objects.

I am, &c.

Yours,

Dudley, April 26th, 1785. J. WAINWRIGHT.

CASE of Mr. WARD, Surgeon, in Birmingham.—Related by himself.

IN *September*, 1782, I was seized with a difficulty of breathing, and oppression in my chest, in consequence of taking cold from being called out in the night. My tongue was foul; my urine small in quantity; my breath laborious and distressing on the slightest exercise. I tried the medicines most generally recommended, such as emetics, blisters, lac ammoniacum, oxymel of squills, &c. but finding little or no relief, I consulted Dr. Withering, who advised me to try the following prescription.

R. Fol.

R. Fol. Digital. purp. ficcat. ʒifs.

 Aq. bullientis ℥iv.

 Aq. cinn. fp. ℥fs. digere per horas quatuor, et colaturæ capiat cochlear. i. nocte maneque.

He alfo defired me to take fifty drops of tincture of cantharides three or four times a day.

After taking eight ounces of the infufion, and about twelve drams of the drops, I was perfectly cured, and have had no return fince. The medicine did not occafion ficknefs or vertigo, nor had they any other fenfible effect than in changing the appearance, and increafing the quantity of the urine, and rendering the tongue clean. After the laft dofe or two indeed, I had a little naufea, which was immediately removed by a fmall glafs of brandy.

Birmingham, 1ft July, 1785.

Communications from Mr. Yonge, Surgeon, in Shiffnall, Shropfhire.

Dear Sir,

   I HAVE great fatisfaction in complying with your juft claim, by tranfcribing outlines of the fubfequent cafes, for infertion in your long requefted tract on the Digitalis purpurea. The two firft of thefe you will eafily recollect, the cures having been conducted immediately under your own manage-

ment

ment, and the whole may add to that weight of evidence which long experience enables you to adduce of the efficacy of that valuable medicine. I have recited the only inftances of its failure which occur to me, but many other, though fuccefsful cafes, wherein its utility might feem dubious, and alfo the accounts received from people whofe accuracy might be fufpected, I fhall not for obvious reafous trouble you with.

I am, dear Sir,

Your obliged friend,

Shiffnall,   WILLIAM YONGE.
May 1, 1785.

C A S E   I.

A Gentleman aged 49, on the night of the 21ft of Auguft, 1784, awaked with a fenfe of fuffocation, which obliged him to rife up fuddenly in bed. I found him complaining of difficult refpiration, particularly on lying down ; the countenance pale, and the pulfe fmaller and quicker than ufual. Some brandy and water having been given, the fymptoms gradually abated, fo that he flept in a half recumbent pofture. The following day he expreffed a fenfe of anxiety and weight in the cheft, attended by quicker breathing upon motion of the body. That evening an emetic of ipecacohana was given, and afterwards a draught, with vitriolic æther

and

and confect. card. aa ʒito be repeated as the symptoms should require it. He continued to be affected with slighter returns of the dyspnœa at irregular intervals, until *September* 15th, when upon a more severe attack, the emetic was repeated. He now recollected some slight pain in his arms which had affected him previous to this last seizure, and was disposed to consider his complaint as rheumatic. Pills with gum ammoniac. gum guaiac. and antimonial powder were directed, with infus. amar. simpl. twice a day. The bowels were regulated by aperient pills of pulv. jalap. aloes and sal. tartar. and ʒiss balsam peruv. was given occasionally to alleviate the paroxysms of dyspnœa.

From this period until the beginning of November, little amendment or variation happened, except that respiration became more permanently difficult, and particularly oppressed upon motion, nor was it relieved by the expectoration of a mucous discharge, which now increased considerably. Squills, musk, ol. succini, æther, with other medicines of the same kind, were now used, but without success. The effects of opium and venæsection were tried. The appetite diminished, and his sleep became short and disturbed. He sometimes slept lying upon his back, but generally upon his left side. The urine which had hitherto been of good colour, and sufficient quantity, now became diminished, and lateritious; and the ancles œdematous.

On

On the 15th of *November* a blifter was laid over the fternum, and ʒifs of oxymel fcillitic. was given every eight hours.

On the 18th, a more copious difcharge of urine took place; the fwelling of the feet foon difappeared, and the refpiration became gradually relieved.

On the 30th ʒi tinɕt. cantharidum twice a day in pyrmont water, with pills of ammoniac, fal tartar. et extraɕt. gentian. were fubftituted, but

On the 7th of *December*, from fome fymptoms of relapfe, the oxymel was ufed as before, and continued to be taken until the 27th, in dofes as large as could be difpenfed with on account of the great naufea which attended fits exhibition: The urine was made in the quantity of four or five pints each day, during the whole time; the quantity then drank being feldom more than three pints. But now the ficknefs being exceedingly deprefling, the ftrength failing, and the diuretic effeɕts beginning to ceafe, the following prefcription was direɕted.

R. Fol. Digitalis purpur. pulv. ɘfs.
    Spec. Aromatic. ɘi. fp. lav. c. f. pilul. no. x. capiat i. noɕte maneque, et alternis diebus fenfim augeatur dofin.

In three days the effeɕt of this medicine became vifible, and when the dofe of the Digitalis had been
                                        increafed

increafed to fix grains per day, the flow of urine generally amounted to feven pints every twenty-four hours. Not the leaft ficknefs, nor any other difagreeable fymptom fupervened, though he per-fevered in this plan until the end of *January* at which time the dyfpnœa was removed, and he has continued gradually to regain his flefh, ftrength, and appetite, without any relapfe.

C A S E   II.

About the middle of the year 1784 a lady aged 48, returned from London, to her native air in Shrop-fhire, under fymptoms of complicated difeafe. It was your opinion that the plethoric ftate, confe-quent to that period, when menftruation firft begins to ceafe, had under various appearances, laid the foundation of that deplorable ftate which now pre-fented itfelf. The fkin was univerfally of a pale, leaden colour; her perfon much emaciated, and her ftrength fo reduced, as to difable her from walking without fupport. The appetite fluctuating, the di-geftion impaired fo much, that folids paffed the inteftines with little appearance of folution : She had generally eight or ten alvine evacuations every day, and without this number, febrile fymptoms, attended with fevere vertiginous affection, and vomiting regularly enfued. The ftools were of a pale afh colour. The urine generally pale, and at firft in due quantity. The region of the ftomach
had

had a tenfe feel, without forenefs : the feet and
ancles œdematous, her fleep was uncertain : the
pulfe varying between 94 and 100, and feeble,
except upon the approach of the menftrual periods,
which were now only marked by its increafed
ftrength, and exacerbation of other febrile fymptoms. Emetics, faline medicines, and gentle ape-
rients were neceffary to alleviate thefe. Six grains
of ipecac. operated with fufficient power, and half
a grain of calomel would have purged with great
violence.

From the time of her arrival till the middle of
Auguft, mercury had been continued in various
forms, and in dofes fuch as the irritable ftate of her
ftomach and bowels would admit of. Spirit. nitri
dulc. ; fal. tartar. fquill, and cantharides were
alternately employed as diuretics, but without fuccefs, to retard the progrefs of an univerfal anafarca,
which was then advanced to fuch degree and accompanied by fo great debility, and other dreadful con-
comitants, as to threaten a fpeedy and fatal cataf-
trophe.

On the 16th of *Auguft* you firft faw her, and
directed thus.

R. Mercur. cinerei gr. ii.
Fol. Digital. purpur. pulv. əi. f. mafs. in
pill. no. xvi. dividend.—fumat unam hora meridia-
ana, iterumque hora quinta pomeridiana quotidie.

Capiat

Capiat lixivii faponac. gutt. **L.** in hauſt. juſcul. ſine ſale parati omni noɕe.

On the 20th the flow of urine began to increaſe, and ſhe continued the medicine in the ſame doſe until the 20th of *September*, diſcharging from ſix to eight pints of water each day for the firſt week, and which quantity gradually diminiſhed as ſhe became empty. During this period ſhe complained not of any ſickneſs, except from the lixivium, which was after the firſt doſe reduced to 20 drops; and her appetite and ſtrength increaſed daily, though it was evident that no bile had yet flowed into the bowels, nor was the digeſtion at all improved. The anaſarcous appearances being then removed, the Digitalis was omitted, and pills, compoſed of mercur. cinereus, aloes, and ſal tartari direɕed twice a day, with 3i. of vin. chalybeat. in infuſ. amar. ſimpl.

Her amendment in other reſpeɕts proceeded ſlowly, but regularly, from that time until the 9th of Oɕtober; when the ſtate of plethora again recurring, with its uſual attendant ſymptoms, ℥iv. of blood were taken from the arm; and this was upon the ſame occaſion, repeated in the following month, with manifeſt good conſequences; though in both inſtances the colour of the blood, as flowing from the vein could hardly be called red, and the coagulum was as weak in its coheſion as poſſible. The ſtate of the ſtomach and bowels was by this time greatly improved, in common with other parts of the

the fyſtem; but no intromiſſion of bile had yet happened: the hardneſs about the hypogaſtric region, though leſs, continued in a conſiderable degree, and you ordered pills of mercury rubbed down, and ruſt of iron, to be taken twice a day, with a decoction of dandelion and ſal ſodæ.

A cataplaſm of linſeed was applied every night over the ſtomach and right ſide; and, with little deviation from this plan, ſhe continued to the end of the year, improving in her general health, but the hepatic affection yet remaining. It was then determined to try the effects of electricity, and gentle ſhocks were paſſed through the body daily, and as nearly as could be through the liver, in various directions.

On the fifth day there was reaſon to think that ſome gall had been ſecreted and poured out, and this became every day more evident; but it flowed only in ſmall quantity, and irregularly into the bowels, as appeared from the fæces being partially tinged by it.

In *February* the lady left this neighbourhood, and though convaleſcent, yet ſo nearly well as to promiſe us the ſatisfaction of ſeeing her perfectly reſtored.

June 29. The bile is now ſecreted in pretty good quantity, her appetite is perfectly good, her ſtrength equal to almoſt any degree of exerciſe, and her
health

health in general better than it has been for fome
years.

C A S E   III.

Mr. W———, aged —. In *June*, 1782, was
affeᴄᴛed with flight difficulty in refpiration, upon taking exercife or lying down in bed. Thefe fymptoms increafed gradually until the end of *July*,
when he complained of fenfe of weight and uneafi-
nefs about the prœcordia; lofs of appetite; and
coftivenefs. The urine was fmall in quantity, and
high coloured; his pulfe feeble, and intermitting;
he breathed with difficulty when in bed, and flept
little. After the exhibition of an emetic, and an
opening medicine of rhubarb, fena, and fal tartari,
he was direᴄᴛed to take half a dram of fquill pill,
pharm. Edinburg. night and morning, with ʒfs fal.
fodæ in ʒifs. infuf. amar. fimpl. twice a day; and
thefe medicines were continued during ten days,
without any fenfible effeᴄᴛ. A blifter was then applied to the fternum, and fix grains of calomel giv-
en in the evening. The fymptoms were now increafed very confiderably, in every particular; and
the following infufion was fubftituted for the former
medicines.

R. Fol. Digital. purpur. ʒiii.

Cort. limon. ʒii. infund.

Aq. bullient. ℔i. per hor. 2 et cola. fumat
cochl. i. primo mane et repet. omni hora.

Sometime

Sometime in the night confiderable naufea occurred, and the following day he began to make water in great quantity, which he continued to do for
three or four days. The pulfe in a few hours became regular, flower, and ftronger, and, in the
courfe of a week, all the fymptoms entirely vanifhed, and an electuary of cort. peruvian, fal martis,
and fpec. aromatic. confirmed his cure.

In *February*, 1784, this gentleman had a relapfe
of his difeafe, from which he again foon recovered
by the fame means, and is now perfectly well.

C A S E   IV.

G—— A——, a hufbandman, aged 57. Was
in the year 1782 affected with a flight, but conftant
pain in his breaft, with difficult refpiration. His
countenance was yellow; the abdomen fwelled, and
hard; his urine high coloured, and in fmall quantity; appetite and fleep little. Complained of frequent naufea, and of fudden profufe fweatings,
which feemed for a fhort time to relieve the dyfpnœa.

After the exhibition of an emetic, fix grains of
calomel were given, with a purge of jalap in the
morning, and repeated in a few days, with fome appearance of advantage. He was then directed to
take fome pills of fquill, foap, and rhubarb, with a
draught twice a day, confifting of infuf. amar. fimp.
and fal tartari. The fkin foon became clearer and
the

the pain in his breaft confiderably diminifhed. But every other circumftance remaining the fame, and a fluctuation in the belly being now more evident, the infufion of Digitalis as prefcribed in cafe third, was given in the dofe of one ounce twice a day.

On the 5th day the effects were apparent, and he continued his medicine for a fortnight without nau-fea, making four or five pints of water every night, but little in the day, and gradually lofing the fymp-toms of his difeafe.

In 1784, this perfon had a relapfe, and was again cured by fimilar treatment.

CASE V.

R——— H———, Aged 43. Towards the end of the year 1783, became affected with flight cough and ex-pectoration of purulent matter. In December his fkin became univerfally of a pale yellow colour. The abdomen was fwelled and hard; his appetite little, and he complained of a violent and conftant palpitation of the heart, which prevented him from fleeping. The urine pale, and in fmall quantity. The pulfe exceedingly ftrong, and rebounding; beating 114 to 120 ftrokes every minute. He fuffer-ed violent pain of his head, and was very feeble and emaciated. After bleeding, and the ufe of gentle aperient medicines, he continued to take the infu-fion of Digitalis for fome days, without any fenfible effect. Other diuretics were tried to as little pur-pofe

pofe. Repeated bleeding had no effect in diminifhing the violent action of the heart. He died in
January following, under complicated fymptoms of
phthifis and afcites.

CASE VI.

A man aged 57, who had lived freely in the fummer of 1784, became affected with œdematous
fwelling of his legs, for which he was advifed to
drink Fox Glove Tea. He took a four ounce bafon
of the infufion made ftrong with the green leaves,
every morning for four fucceffive days.

On the 5th he was fuddenly feized with faintnefs
and cold fweatings. I found him with a pale countenance, complaining of weaknefs, and of pain,
with a fenfe of great heat in his ftomach and
bowels. The fwelling of the legs was entirely gone,
he having evacuated urine in very large quantities
for the two preceding days. He was affected with
frequent diarrhœa. The pulfe was very quick and
fmall, and his extremities cold.

A fmall quantity of broth was directed to be given
him every half hour, and blifters were applied
to the ancles, by which his fymptoms became gradually alleviated, and he recovered perfectly in the
fpace of three weeks; except a relapfe of the anafarca, for which the Digitalis was afterwards fuccefs-
fully employed, in fmall dofes, without any difagreeable confequence.

CASE

C A S E   VII.

S—— D————, a middle aged single woman, was affected in the year eighty-one, with a painful rigidity and slight inflammation of the integuments on the left side, extending from the ear to the shoulder. In every other particular she was healthy. The use of warm fomentations, and opium, with two or three doses of mercurial physic, afforded her ease and the inflammation disappeared, but was succeeded by an œdematous swelling of the part, which very gradually extended along the arm, and downward to the breast, back, and belly. Friction, electricity and mercurial ointment were amongst the number of applications unsuccessfully employed to relieve her for the space of three months, during which time she continued in good general health.

In *November* she became ascitic, passing small quantities of urine, and soon afterwards a sudden dyspnœa gave occasion to suppose an effusion of water in the thorax. The Digitalis, squills, and cantharides were given in very considerable doses without effect. She died the latter end of December following.

C A S E   VIII.

W—— C————, a collier aged 58, was attacked in the spring of 1783 with a tertian ague, which he attributed to cold, by sleeping in a coal

M   pit,

pit, and from which he recovered in a few days, except a swelling of the lower extremities, which had appeared about that time, and gradually increased for two or three months. The legs and thighs were greatly enlarged and œdematous. His belly was swelled, but no fluctuation perceptible. He made small quantities of high coloured water. The appetite bad, and pulse feeble. He had taken many medicines without relief, and was now so reduced in strength, as to sit up with difficulty. An infusion of the Digitalis was directed for him, in the proportion of one ounce of the fresh leaves to a pint of water, two ounces to be taken three times a day, until the stomach or bowels became affected. Upon the exhibition of the sixth dose, nausea supervened, and continued to oppress him at intervals for two or three days, during which he passed large quantities of pale urine. The swelling, assisted by moderate bandage rapidly diminished, and without any repetition of his medicine, at the expiration of sixteen days, he returned to his labour perfectly recovered.

O F

OF THE

PREPARATIONS and DOSES,

OF THE

FOXGLOVE.

EVERY part of the plant has more or lefs of the fame bitter tafte, varying, however, as to ftrength, and changing with the age of the plant and the feafon of the year.

ROOT.—This varies greatly with the age of the plant. When the ftem has fhot up for flowering, which it does the fecond year of its growth, the root becomes dry, nearly taftelefs, and inert.

Some practitioners, who have ufed the root, and been fo happy as to cure their patients without exciting ficknefs, have been pleafed to communicate the circumftance to me as an improvement in the ufe of the plant. I have no doubt of the truth of their remarks, and I thank them. But the cafe of Dr. Cawley puts this matter beyond difpute. The fact is, they have fortunately happened to ufe the root in its approach to its inert ftate, and confequently have not over dofed their patients. I could,

M 2

if

if neceſſary, bring other proof to ſhew that the root
is juſt as capable as the leaves, of exciting nauſea.

S T E M.—The ſtem has more taſte than the root
has, in the ſeaſon the ſtem ſhoots out, and leſs taſte
than the leaves.   I do not know that it has been
particularly ſelected for uſe.

L E A V E S. —Theſe vary greatly in their efficacy at different ſeaſons of the year, and, perhaps,
at different ſtages of their growth; but I am not
certain that this variation keeps pace with the greater
or leſſer intenſity of their bitter taſte.

Some who have been habituated to the uſe of
the recent leaves, tell me, that they anſwer their
purpoſe at every ſeaſon of the year; and I believe
them, notwithſtanding I myſelf have found very
great variations in this reſpect.   The ſolution of
this difficulty is obvious. They have uſed the leaves
in ſuch large proportion, that the doſes have been
ſufficient, or more than ſufficient, even in their
moſt inefficacious ſtate.   *The Leaf-ſtalks* ſeem, in
their ſenſible properties, to partake of an interme-
diate ſtate between the leaves and the ſtem.

F L O W E R S.—The petals, the chives, and the
pointal have nearly the taſte of the leaves, and it
has been ſuggeſted to me, by a very ſenſible and
judicious friend, that it might be well to fix on the
flower for internal uſe.   I ſee no objection to the
propoſition; but I have not tried it.

SEEDS.

S E E D S.—Thefe I believe are equally untried.

From this view of the different parts of the plant, it is fufficiently obvious why I ftill continue to pre-fer the leaves.

Thefe fhould be gathered after the flowering ftem has fhot up, and about the time that the bloffoms are coming forth.

The leaf-ftalk and mid-rib of the leaves fhould be rejected, and the remaining part fhould be dried, either in the fun-fhine, or on a tin pan or pewter difh before a fire.

If well dried, they readily rub down to a beauti-ful green powder, which weighs fomething lefs than one-fifth of the original weight of the leaves. Care muft be taken that the leaves be not fcorched in drying, and they fhould not be dried more than what is requifite to allow of their being readily re-duced to powder.

I give to adults, from one to three grains of this powder twice a day. In the reduced ftate in which phyficians generally find dropfical patients, four grains a day are fufficient. I fometimes give the powder alone; fometimes unite it with aromatics, and fometimes form it into pills with a fufficient quantity of foap or gum ammoniac.

If

If a liquid medicine be preferred, I order a dram of thefe dried leaves to be infufed for four hours in half a pint of boiling water, adding to the ftrained liquor an ounce of any fpirituous water. One ounce of this infufion given twice a day, is a medium dofe for an adult patient. If the patient be ftronger than ufual, or the fymptoms very urgent, this dofe may be given once in eight hours ; and on the contrary in many inftances half an ounce at a time will be quite fufficient. About thirty grains of the powder or eight ounces of the infufion, may generally be taken before the naufea commences.

The ingenuity of man has ever been fond of ex-erting itfelf to vary the forms and combinations of medicines. Hence we have fpirituous, vinous, and acetous tinctures ; extracts hard and foft, fyrups with fugar or honey, &c. but the more we multi-ply the forms of any medicine, the longer we fhall be in afcertaining its real dofe. I have no lafting objection however to any of thefe formulæ except the extract, which, from the nature of its prepara-tion muft ever be uncertain in its effects ; and a medicine whofe fulleft dofe in fubftance does not exceed three grains, cannot be fuppofed to ftand in need of condenfation.

It appears from feveral of the cafes, that when the Digitalis is difpofed to purge, opium may be joined with it advantageoufly ; and when the bowels are too tardy, jalap may be given at the fame time, without

without interfering with its diuretic effects; but I have not found benefit from any other adjunct.

From this view of the doses in which the Digitalis really ought to be exhibited, and from the evidence of many of the cases, in which it appears to have been given in quantities fix, eight, ten or even twelve times more than neceffary, we muft admit as an inference either that this medicine is perfectly fafe when given as I advife, or that the medicines in daily ufe are highly dangerous.

EFFECTS,

EFFECTS, RULES, and CAUTIONS.

———————

THE Foxglove when given in very large and quick-ly-repeated dofes, occafions ficknefs, vomiting, purging, giddinefs, confufed vifion, objects appearing green or yellow ; increafed fecretion of urine, with frequent motions to part with it, and fometimes inability to retain it ; flow pulfe, even as flow as 35 in a minute, cold fweats, convulfions, fyncope, death.*

When given in a lefs violent manner, it produces moft of thefe effects in a lower degree ; and it is curious to obferve, that the ficknefs, with a certain dofe of the medicine, does not take place for many hours after its exhibition has been difcontinued ; that the flow of urine will often precede, fometimes accompany, frequently follow the ficknefs at the diftance of fome days, and not unfrequently be checked by it. The ficknefs thus excited, is extremely different from that occafioned by any other medicine ; it is peculiarly diftreffing to the patient ; it ceafes, it recurs again as violent as before ; and thus it will continue to recur for three or four days, at diftant and more diftant intervals.

Thefe

———————

* I am doubtful whether it does not fometimes excite a copious flow of faliva.—See cafes at pages 115, 154, and 155.

Thefe fufferings of the patient are generally rewarded by a return of appetite, much greater than what exifted before the taking of the medicine.

But thefe fufferings are not at all neceffary; they are the effects of our inexperience, and would in fimilar circumftances, more or lefs attend the exhibition of almoft every active and powerful medicine we ufe.

Perhaps the reader will better underftand how it ought to be given, from the following detail of my own improvement, than from precepts peremptorily delivered, and their fource veiled in obfcurity.

At firft I thought it neceffary *to bring on and continue the ficknefs, in order to enfure the diuretic effects.*

I foon learnt that the naufea being once excited, it was unneceffary to repeat the medicine, as it was certain to recur frequently, at intervals more or lefs diftant.

Therefore my patients were ordered *to perfift until the naufea came on, and then to ftop.* But it foon appeared that the diuretic effects would often take place firft, and fometimes be checked when the ficknefs or a purging fupervened.

The

The direction was therefore enlarged thus—*Continue the medicine until the urine flows, or sickness or purging take place.*

I found myself safe under this regulation for two or three years ; but at length cases occurred in which the pulse would be retarded to an alarming degree, without any other preceding effect.

The directions therefore required an additional attention to the state of the pulse, and it was moreover of consequence not to repeat the doses too quickly, but to allow sufficient time for the effects of each to take place, as it was found very possible to pour in an injurious quantity of the medicine, before any of the signals for forbearance appeared.

Let the medicine therefore be given in the doses, and at the intervals mentioned above:—let it be continued until it either acts on the kidneys, the stomach, the pulse, or the bowels; let it be stopped upon the first appearance of any one of these effects, and I will maintain that the patient will not suffer from its exhibition, nor the practitioner be disappointed in any reasonable expectation.

If it purges, it seldom succeeds well.

The patients should be enjoined to drink very freely during its operation. I mean, they should drink whatever they prefer, and in as great quantity

tity as their appetite for drink demands.   This direction is the more neceffary, as they are very generally prepoffeffed with an idea of drying up a
dropfy, by abftinence from liquids, and fear to add
to the difeafe, by indulging their inclination to
drink.

In cafes of afcites and anafarca ; when the patients are weak, and the evacuation of the water
rapid ; the ufe of proper bandage is indifpenfably
neceffary to their fafety.

If the water fhould not be wholly evacuated,
it is beft to allow an interval of feveral days before
the medicine be repeated, that food and tonics may
be adminiftered ; but truth compels me to fay, that
the ufual tonic medicines have in thefe cafes very
often deceived my expectations.

From fome cafes which have occurred in the
courfe of the prefent year, I am difpofed to believe
that the Digitalis may be given in fmall dofes, viz.
two or three grains a day, fo as gradually to remove
a dropfy, without any other than mild diuretic effects, and without any interruption to its ufe until
the cure be compleated.

If inadvertently the dofes of the Foxglove fhould
be prefcribed too largely, exhibited too rapidly, or
urged to too great a length ; the knowledge of a
remedy to counteract its effects would be a defirable
thing.

thing.   Such a remedy may perhaps in time be difcovered.   The ufual cordials and volatiles are generally rejected from the ftomach ; aromatics and ftrong bitters are longer retained ; brandy will fometimes remove the ficknefs when only flight ; I have fometimes thought fmall dofes of opium ufeful, but I am more confident of the advantage from blifters. Mr. Jones *(Page* 135) in one cafe, found mint tea to be retained longer than other things.

C O N-

CONSTITUTION of PATIENTS.

INDEPENDENT of the degree of difeafe, or of the ftrength or age of the patient, I have had occafion to remark, that there are certain conftitutions favourable, and others unfavourable to the fuccefs of the Digitalis.

From large experience, and attentive obfervation, I am pretty well enabled to decide *a priori* upon this matter, and I wifh to enable others to do the fame: but I feel myfelf hardly equal to the undertaking. The following hints, however, aiding a degree of experience in others, may lead them to accomplifh what I yet can defcribe but imperfectly.

It feldom fucceeds in men of great natural ftrength, of tenfe fibre, of warm fkin, of florid complexion, or in thofe with a tight and cordy pulfe.

If the belly in afcites be tenfe, hard, and circumfcribed, or the limbs in anafarca folid and refifting, we have but little to hope.

On the contrary, if the pulfe be feeble or intermitting, the countenance pale, the lips livid, the fkin cold, the fwollen belly foft and fluctuating, or

the

the anafarcous limbs readily pitting under the preffure of the finger, we may expect the diuretic effects to follow in a kindly manner.

In cafes which foil every attempt at relief, I have
been aiming, for fome time paft, to make fuch a
change in the.conftitution of the patient, as might
give a chance of fuccefs to the Digitalis.

By blood-letting, by neutral falts, by chryftals
of tartar, fquills, and occafional purging, I have
fucceeded, though imperfectly. Next to the ufe
of the lancet, I think nothing lowers the tone of
the fyftem more effectually than the fquill, and confequently it will always be proper, in fuch cafes, to
ufe the fquill; for if that fail in its defired effect, it
is one of the beft preparatives to the adoption of the
Digitalis.

A tendency to paralytic affections, or a ftroke of
the palfy having actually taken place, is no objection to the ufe of the Digitalis; neither does a
ftone exifting in the bladder forbid its ufe. Theo-
retical ideas of fedative effects in the former, and
apprehenfions of its excitement of the urinary or-
gans in the latter cafe, might operate fo as to
make us with-hold relief from the patient; but experience tells me, that fuch apprehenfions are
groundlefs.

INFER-

INFERENCES.

TO prevent any improper influence, which the above recitals of the efficacy of the medicine, aided by the novelty of the fubject, may have upon the minds of the younger part of my readers, in raifing their expectations to too high a pitch, I beg leave to deduce a few inferences, which I apprehend the facts will fairly fupport.

I. That the Digitalis will not univerfally act as a diuretic.

II. That it does do fo more generally than any other medicine.

III. That it will often produce this effect after every other probable method has been fruitlefsly tried.

IV. That if this fails, there is but little chance of any other medicine fucceeding.

V. That in proper dofes, and under the management now pointed out, it is mild in its operation, and gives lefs difturbance to the fyftem, than fquill, or almoft any other active medicine.

VI. That when dropfy is attended by palfy, un-found vifcera, great debility, or other complication of difeafe, neither the Digitalis, nor any other diuretic

retic can do more than obtain a truce to the urgency of the fymptoms; unlefs by gaining time, it may
afford opportunity for other medicines to combat
and fubdue the original difeafe.

VII. That the Digitalis may be ufed with advantage in every fpecies of dropfy, except the encyfted.

VIII. That it may be made fubfervient to the
cure of difeafes, unconnected with dropfy.

IX. That it has a power over the motion of the
heart, to a degree yet unobferved in any other medicine, and that this power may be converted to falutary ends.

PRACTICAL

PRACTICAL

REMARKS ON DROPSY,

AND SOME OTHER DISEASES.

THE following remarks confift partly of matter of fact, and partly of opinion. The former will be permanent; the latter muft vary with the detection of error, or the improvement of knowledge. I hazard them with diffidence, and hope they will be examined with candour; not by a contraft with other opinions, but by an attentive comparifon with the phœnomena of difeafe.

ANASARCA.

§ 1. THE anafarca is generally curable when feated in the fub-cutaneous cellular membrane, or in the fubftance of the lungs.

§ 2. When the abdominal vifcera in general are greatly enlarged, which they fometimes are, without effufed fluid in the cavity of the abdomen; the difeafe is incurable. After death, the more folid vifcera are found very large and pale. If the cavity contains water, that water may be removed by diuretics.

N   § 3. In

§ 3. In fwollen legs and thighs, where the refift-ance to preffure is confiderable, the tendency to tranfparency in the fkin not obvious, and where the alteration of pofture occafions but little alteration in the ftate of diftenfion, the cure cannot be effect-ed by diuretics.

Is this difficulty of cure occafioned by fpiffitude in the effufed fluids, by want of proper communi-cation from cell to cell, or is the difeafe rather caufed by a morbid growth of the folids, than by an accu-mulation of fluid?

Is not this difeafe in the limbs fimilar to that of the vifcera (§ 2)?

§ 4. Anafarcous fwellings often take place in pal-fied limbs, in arms as well as legs; fo that the fwel-ling does not depend merely upon pofition.

§ 5. Is there not caufe to fufpect that many drop-fies originate from paralytic affections of the lym-phatic abforbents? And if fo, is it not probable that the Digitalis, which is fo effectual in removing dropfy, may alfo be ufed advantageoufly in fome kinds of palfy?

A S C I T E S,

§ 6. IF exifting alone, (i. e.) without accompa-nying anafarca, is in children curable; in adults generally incurable by medicines. Tapping may be ufed

ufed here with better chance for fuccefs than in
more complicated dropfies. Sometimes cured by
vomiting.

ASCITES and ANASARCA.

§ 7. I N C U R A B L E if dependant upon
irremediably difeafed vifcera, or on a gouty conftitution, fo debilitated, that the gouty paroxyfms no
longer continue to be formed.

In every other fituation the difeafe yields to diuretics and tonics.

A S C I T E S, A N A S A R C A, and H Y D R O T H O R A X.

§ 8. U N D E R this complication, though the
fymptoms admit of relief, the reftoration of the
conftitution can hardly be hoped for.

A S T H M A.

§ 9. T H E true fpafmodic afthma, a rare difeafe
—is not relieved by Digitalis.

§ 10. In the greater part of what are called
afthmatical cafes, the real difeafe is anafarca of the
lungs, and is generally to be cured by diuretics. (See
§ 1.) This is almoft always combined with fome
fwelling of the legs.

§ 11. There

§ 11. There is another kind of afthma, in which change of pofture does not much affect the patient. I believe it to be caufed by an infarction of the lungs. It is incurable by diuretics; but it is often accompanied with a degree of anafarca, and fo far it admits of relief.

Is not this difeafe fimilar to that in the limbs at (§ 3,) and alfo to that of the abdominal vifcera at (§ 2.)?

ASTHMA and ANASARCA.

§ 12. IF the afthma be of the kind mentioned at (§§ 9 and 11,) diuretics can only remove the accompanying anafarca. But if the affection of the breath depends alfo upon cellular effufion, as it moftly does, the patient may be taught to expect a recovery.

ASTHMA and ASCITES.

§ 13. A RARE combination, but not incurable if the the abdominal vifcera are found. The afthma is here moft probably of the anafarcous kind (§ 10;) and this being feldom confined to the lungs only, the difeafe generally appears in the following form.

ASTHMA,

ASTHMA, ASCITES, and ANASARCA.

§ 14. THE curability of this combination will depend upon the circumftances mentioned in the preceding fection, taking alfo into the account the ftrength or weaknefs of the patient.

E P I L E P S Y.

§ 15. I N epilepfy dependant upon effufion, the Digitalis will effect a cure; and in the cafes alluded to, the dropfical fymptoms were unequivocal. It has not had a fufficient trial in my hands, to determine what it can do in other kinds of epilepfy,

HYDATID DROPSY.

§ 16. THIS may be diftinguifhed from common afcites, by the want of evident fluctuation. It is common to both fexes. It does not admit of a cure either by tapping or by medicine.

HYDROCEPHALUS.

§ 17. THIS difeafe, which has of late fo much attracted the attention of the medical world, I believe, originates in inflammation; and that the water found in the ventricles of the brain after death, is the confequence, and not the caufe of the illnefs.

It has feldom happened to me to be called upon in the earlier ftages of this complaint, and the fymp-

N 3   toms

toms are at firſt ſo ſimilar to thoſe uſually attendant upon dentition and worms, that it is very difficult to pronounce decidedly upon the real nature of the diſeaſe; and it is rather from the failure of the uſual modes of relief, than from any other more decided obſervation, that we at length dare to give it a name.

At firſt, the febrile ſymptoms are ſometimes ſo unſteady, that I have known them miſtaken for the ſymptoms of an intermittent, and the cure attempt-ed by the bark.

In the more advanced ſtages, the diagnoſtics ob-trude themſelves upon our notice, and put the ſitu-ation of the patient beyond a doubt. But this does not always happen. The variations of the pulſe, ſo accurately deſcribed by the late Dr. Whytt, do not always enſue. The dilatation of the pupils, the ſquinting, and the averſion to light, do not univerſally exiſt. The ſcreaming upon raiſing the head from the pillow or the lap, and the fluſhing of the cheeks, I once conſidered as affording indubita-ble marks of the diſeaſe; but in a child which I ſometime ſince attended with Dr. Aſh, the pulſe was uniformly about 85, (except during the firſt week, before we had the care of the patient.) The child never ſhewed any averſion to the light; never had dilated pupils, never ſquinted, never ſcreamed when raiſed from the lap or taken out of the bed, nor did we obſerve any remarkable fluſhing of the cheeks; and the ſleep was quiet, but ſometimes moaning.

Frequent

Frequent vomiting exifted from the firft, but ceafed for feveral days towards the conclufion. One or two worms came away during the illnefs, and it was all along difficult to purge the child. Three days before death, the right fide became flightly paralytic, and the pupil of that eye fomewhat dilated.

After death, about two ounces and a half of water were found in the ventricles of the brain, and the veffels of the dura mater were turgid with blood.

If I am right as to the nature of hydrocephalus, that it is at firft dependant upon inflammation, or congeftion; and that the water in the ventricles is a confequence, and not a caufe of the difeafe; the curative intentions ought to be extremely different in the firft and the laft ftages.

It happens very rarely that I am called to patients at the beginning, but in two inftances wherein I was called at firft, the patients were cured by repeated topical bleedings, vomits, and purges.

Some years ago I mentioned thefe opinions, and the fuccefs of the practice refulting from them, to Dr. Quin, now phyfician at Dublin. That gentleman had lately taken his degree, and had chofen hydrocephalus for the fubject of his thefis in the year 1779. In this very ingenious effay, which he gave me the fame morning, I was much pleafed to find that the author had not only held the fame

ideas

ideas relative to the nature of the difeafe, but had alfo confirmed them by diffections.

In the year 1781, another cafe in the firft ftage demanded my attention. The reader is referred back to Cafe LXIX for the particulars.

I have not yet been able to determine whether the Digitalis can or cannot be ufed with advantage in the fecond ftage of the hydrocephalus. In Cafe XXXIII. the fymptoms of death were at hand; in Cafe LXIX. the practice, though fuccefsful, was too complicated, and in Cafe CLI. the medicine was certainly ftopped too foon.

When we confider what enormous quantities of mercury may be ufed in this complaint, without af- fecting the falivary glands, it feems probable that other parts may be equally infenfible to the action of their peculiar ftimuli, and therefore that the Di- gitalis ought to be given in much larger dofes in this, than in other difeafes.

HYDROTHORAX.

§ 18. UNDER this name I alfo include the dropfy of the pericardium.

The intermitting pulfe, and pain in the arms, fuf- ficiently diftinguifh this difeafe from afthma, and and from anafarcous lungs.

It is very univerfally cured by the Digitalis.

§ 19, I lately

§ 19. I lately met with two cases which had been considered and treated as angina pectoris. They both appeared to me to be cases of hydrothorax. One subject was a clergyman, whose strength had been so compleatly exhausted by the continuance of the disease, and the attempts to relieve it, that he did not survive many days. The other was a lady, whose time of life made me suspect effusion. I directed her to take small doses of the pulv. Digitalis, which in eight days removed all her complaints. This happened six months ago, and she remains perfectly well.

HYDROTHORAX and ANASARCA.

§ 20. THIS combination is very frequent, and, I believe, may always be cured by the Digitalis.

§ 21. Dropsies in the chest either with or without anasarcous limbs, are much more curable than those of the belly. Probably because the abdominal viscera are more frequently diseased in the latter than in the former cases.

I N S A N I T Y.

§ 22. I APPREHEND this disease to be more frequently connected with serous effusion than has been commonly imagined.

§ 23. Where appearances of anasarca point out the true cause of the complaint, as in cases XXIV. and
XXXIV.

XXXIV. the happieft effects may be expected from
the Digitalis ; and men of more experience than myfelf in cafes of infanity, will probably employ it fuc-
cefsfully in other lefs obvious circumftances.

NEPHRITIS CALCULOSA.

§ 24. WE have had fufficient evidence of the efficacy of the Foxglove in removing the Dyfuria and
other fymptoms of this difeafe ; but probably it is
not in thefe cafes preferable to the tobacco.*

OVARIUM DROPSY.

§ 25. THIS fpecies of encyfted dropfy is not without difficulty diftinguifhable from an afcites ; and
yet it is neceffary to diftinguifh them, becaufe the
two difeafes require different treatment and becaufe
the probality of a cure is much greater in one than
in the other.

§ 26. The ovarium dropfy is generally flow in its
progrefs ; for a confiderable time the patient though
fomewhat emaciated, does not lofe the appearance
of health, and the urine flows in the ufual quantity.
It is feldom that the practitioner is called in early
enough to diftinguifh by the feel on which fide the
cyft originated, and the patients do not attend to
that circumftance themfelves. They generally men-
ftruate

---

* See an original and valuable treatife by Dr. Fowler, entitled,
Medical Reports of the Effects of Tobacco.

ftruate regularly in the incipient ftate of the difeafe, and it is not until the preffure from the fac becomes very great, that the urinary fecretion diminifhes. In this fpecies of dropfy, the patients, upon being queftioned, acknowledge even from a pretty early date, pains in the upper and inner parts of the thighs, fimilar to thofe which women experience in a ftate of pregnancy. Thefe pains are for a length of time greater in one thigh than in the other, and I believe it will be found that the difeafe originated on that fide.

§ 27. The ovarium dropfy defies the power of medicine. It admits of relief, and fometimes of a cure, by tapping. I fubmit to the confideration of practitioners, how far we may hope to cure this difeafe by a feton or a cauftic. —— In the LXIft cafe the patient was too much reduced, and the difeafe too far advanced to allow of a cure by any method; but it teaches us that a cauftic may be ufed with fafety.

§ 28. When tapping becomes neceffary, I always advife the adoption of the waiftcoat bandage or belt, invented by the late very juftly celebrated Dr. Monro, and defcribed in the firft volume of the Medical Effays. I alfo enjoin my patients to wear this bandage afterwards, from a perfuafion that it retards the return of the difeafe. The proper ufe of bandage, when the diforder firft difcovers itfelf, certainly contributes much to prevent its increafe.

O V A-

OVARIUM DROPSY with ANASARCA.

§ 29. THE anafarca does not appear until the encyſted dropſy is very far advanced. It is then probably cauſed by weakneſs and preſſure. The Digitalis removes it for a time.

PHTHISIS PULMONALIS.

§ 30. This is a very increaſing malady in the preſent day. It is no longer limited to the middle part of life : children at five years of age die of it, and old people at ſixty or ſeventy. It is not confined to the flat-cheſted, the fair-ſkinned, the blue eyed, the light-haired, or the ſcrophulous : it often attacks people with full cheſts, brown ſkins, dark hair and eyes, and thoſe in whoſe family no ſcrophulous taint can be traced. It is certainly infectious. The very ſtrict laws ſtill exiſting in Italy to prevent the infection from conſumptive patients, were probably not enacted originally without a ſufficient cauſe. We ſeem to be approaching to that ſtate which firſt made ſuch reſtrictions neceſſary, and in the further courſe of time, the diſeaſe will probably fall off again, both in virulency and frequency.

§ 31. The younger part of the female ſex are liable to a diſeaſe very much reſembling a true conſumption, and from which it is difficult to diſtinguiſh it ; but this diſeaſe is curable by ſteel and bitters. A criterion of true phthiſis has been ſought for in the

ſtate

ftate of the teeth ; but thè exceptions to that rule
are numerous. An unufual dilatation of the pupil
of the eye, is the moſt certain charaċteriſtic.*

§ 32. Sydenham afferts, that the bark did not
more certainly cure an intermittent, than riding did
a confumption.   We muſt not deny the truth of an
affertion, from fuch authority, but we muſt conclude
that the difeafe was more eafily curable a century
ago than it is at prefent.

§ 33. If the Digitalis is no longer ufeful in confumptive cafes, it muſt be that I know not how to
manage it, or that the difeafe is more fatal than formerly ; for it would be hard to deny the teſtimony
cited at page 9.   I wiſh others would undertake
the enquiry.

§ 34. When phthifis is accompanied with anafarca,
or when there is reafon to fufpeċt hydrothorax, the
Digitalis will often relieve the fufferings, and pro-
long the life of the patient.

§ 35. Many

---

* Many years ago I communicated to my friend, Dr. Percival,
an account of fome trials of breathing fixed air in confumptive
cafes.   The refults were publiſhed by him in the fecond Vol. of
his very ufeful Effays Medical and Experimental, and have fince
been copied into other publications.   I take this opportunity of
acknowledging that I fufpeċt myfelf to have been miſtaken in the
nature of the difeafe there mentioned to have been cured.   I believe it was a cafe of *Vomica*, and not a true *Phthifis* that was cured.
The Vomica is almoſt always curable.   The fixed air correċts the
fmell of the matter, and very ſhortly removes the heċtic fever.
My patients not only infpire it, but I keep large jars of the effer-
vefcing mixture conſtantly at work in their chambers.

§ 35. Many years ago, during an attendance upon Mr. B——, of a confumptive family, and himfelf in the laft ftage of a phthifis; after he was fo ill as to be confined to his chamber, his breathing became fo extremely difficult and diftreffing, that he wifhed rather to die than to live, and urged me warmly to devife fome mode to relieve him. Sufpecting ferous effufion to be the caufe of this fymptom, and he being a man of fenfe and refolution, I fully explained my ideas to him, and told him what kind of operation might afford him a chance of relief; for I was then but little acquainted with the Digitalis. He was earneft for the operation to be tried, and with the affiftance of Mr. Parrott, a very refpectable furgeon of this place, I got an opening made between the ribs upon the lower and hinder part of the thorax. About a pint of fluid was immediately difcharged, and his breath became eafy. This fluid coagulated by heat.

After fome days a copious purulent difcharge iffued from the opening, his cough became lefs troublefome, his expectoration lefs copious, his appetite and ftrength returned, he got abroad, and the wound, which became very troublefome, was allowed to heal.

He then undertook a journey to London; whilft there he became worfe: returned home, and died confumptive fome weeks afterwards.

PUER-

P U E R P E R A L   A N A S A R C A.

§ 36. THIS difeafe admits of an eafy and certain cure by the Digitalis.

§ 37. This fpecies of dropfy may originate from other caufes than child birth. In the beginning of laft *March*, a gentleman at Wolverhampton defired my advice for very large and painful fwelled legs and thighs. He was a temperate man, not of a dropfical habit, had great pain in his groins, and attributed his complaints to a fall from his horfe. He had taken diuretics, and the ftrongeft draftic purgatives with very little benefit. Confidering the anafarca as caufed by the difeafed inguinal glands, I ordered common poultice and mercurial ointment to the groins, three grains of pulv. fol. Digitalis night and morning, and a cooling diuretic decoction in the day-time. He foon loft his pain, and the fwellings gradually fubfided.

T H E   E N D.

Printed in the United States
By Bookmasters